Ernst Schering Research Foundation Workshop 59
Appropriate Dose Selection –
How to Optimize Clinical Drug Development

Wolfgang Meyer 09/08

Ernst Schering Research Foundation
Workshop 59

Appropriate Dose Selection – How to Optimize Clinical Drug Development

J. Venitz, W. Sittner
Editors

With 36 Figures

Springer

Series Editors: G. Stock and M. Lessl

Library of Congress Control Number: 2006928310

ISSN 0947-6075

ISBN-10 3-540-27867-2 Springer Berlin Heidelberg New York
ISBN-13 978-3-540-27867-2 Springer Berlin Heidelberg New York

Springer is a part of Springer Science+Business Media
springer.com

Editor: Dr. Ute Heilmann, Heidelberg
Desk Editor: Wilma McHugh, Heidelberg
Production Editor: Monika Riepl, Leipzig
Cover design: WMXDesign GmbH, Heidelberg
Typesetting and production: LE-TEX Jelonek, Schmidt & Vöckler GbR, Leipzig
21/3100/YL – 5 4 3 2 1 0 Printed on acid-free paper

Preface

Cancer has become a chronic disease, often requiring long-term, chronic oncological drug treatment. As a result, the oncological treatment is exposed to a large number of patients who may be on concurrent treatments for other health conditions and/or who may suffer from concomitant illnesses, both of which may affect the efficacy and safety of the oncological treatment by contributing to the increased incidence of adverse events and/or loss of efficacy.

Newer oncological drugs also have become more targeted to the underlying disease process(es), and their use and dosage regimens may need to be tailored to individual patients. Therefore, as opposed to the older chemotherapeutic drugs, these new agents are usually not dosed to their maximal tolerated dose (MTD), and optimal dose individualization has become more important to improve their clinical efficacy and safety.

At the same time, the overall drug development process has become more time-consuming and expensive while more potential biological targets in cancer are being explored. Novel drug candidates are screened in early clinical drug development (ECDD), based on their clinical safety, pharmacokinetic (PK) properties and achievement of desired biological effects; drug candidates surviving this early clinical screen undergo more rigorous and large-scale phase III testing to demonstrate their safety and efficacy for regulatory approval and clinical use.

Unfortunately, quite a few of these new oncological agents, both approved and under development, have less favorable biopharmaceutical characteristics, especially for chronic oral administration, e.g., poor gastrointestinal solubility and/or permeability and/or extensive drug metabolism, leading to low oral bioavailability and a high degree of variability among patients in systemic drug exposure and pharmacodynamic (PD) or clinical drug response. Effective chronic cancer treatment requires information on intrinsic and extrinsic patient factors impacting the clinical drug response, e.g., drug–drug interactions (DDIs), pharmacogenetic (PG) differences, etc. The basic sciences provide more, albeit incomplete, knowledge about the fundamental mechanisms of drug disposition, such as drug transporters and metabolic pathways as well as disease biology, such as biological targets, pathophysiological pathways, and intermediate markers, i.e., potential biomarkers (BMs).

This increasing body of knowledge is only slowly translated into the optimal clinical use of these agents:

- We appreciate the role that patient subpopulations may play in the outcomes of clinical trials, based on their response or lack thereof to drug treatment.
- We are learning more about possible BMs of drug response, efficacy, and safety, which may assist in optimal dose finding.

- We may have to carefully individualize dosage regimens for individual patients.

Optimal dose finding requires integration of exposure–response (ER) relationships throughout the clinical drug development (CDD) process by quantitative methods. Frontloading CDD to fail poor drug candidates early in the development process has put the burden on early clinical drug development to:

- Provide early proof of (biological and/or therapeutic) concept (POC) for novel targets/drug candidates in order to make go/no go decisions.
- Rationally select dosing regimens in phase I and II information to optimize chances of success in phase III by integrating information across in vitro/in vivo (different species) PK/PD/clinical studies to select appropriate doses and BM.
- Support drug product labeling information for safe and effective use in postapproval clinical use.

Ultimately, the clinical development process has to provide the following information for the optimal clinical use in order to minimize risk and maximize benefit of the new drug treatment:

1. What are the odds of achieving desired clinical outcomes, i.e., efficacy without toxicity?
   Which patients are at high or low risk or stand to benefit the most? Can these subpopulations be identified a priori? How should the dosage regimen be individualized to minimize harm and maximize benefit? What marker or level of exposure can be monitored during long-term treatment, and how should the dosage regimen be adjusted, if necessary, e.g., due to DDI?
2. What are the stakes in terms of benefit or harm?
   What are the clinical consequences of lack of efficacy and adverse events, given the seriousness of the underlying cancer disease? What are clinical alternative dosing regimens, drug combinations, or other treatments?

The answers to these questions, based on empiric evidence, theoretical considerations (e.g., PK/PD modeling), and clinical judgment are mandatory prerequisites for rational drug development and risk management by drug developers and manufacturers, regulatory agencies, health care providers, and society at large.

This workshop brought together experts from throughout the world to present and discuss the following issues:

1. How to demonstrate POC (biological/therapeutic) in ECDD and how to design pivotal phase I/II programs.
2. How to evaluate BM in ECCD and throughout CDD.
3. How to explore ER for drug response and toxicity throughout CDD and how to design optimal dose ranging studies in phases I and II.
4. How to select an appropriate, i.e., likely to succeed in phase III, dosing regimen in ECDD and how to define "optimal dose".
5. How to label a drug product for safe and effective clinical use after approval in the marketplace.

After excellent presentations and frank discussions, consensus emerged that the current preponderance of *empiricism* in oncology drug development should be replaced by a better *mechanistic* understanding of the underlying disease biology and target along with the human drug disposition for the newer agents, especially the targeted drugs. Co-developing BM with the drug candidate early can be extremely helpful in early POC and better dose selection. This may also help to fail early and cheaply drug candidates with a low likelihood to succeed in phase III. In addition, better dose–response trials (e.g., randomized phase II trials) are needed, incorporating BM and/or clinical response; however, the concept of the optimal biologically effective dose has been questioned recently due to our knowledge gap in disease biology. Given the increasing polypharmacy and co-morbidities in cancer patients, it has become essential to identify important clinical covariates, leading to PK and PD variability among patients. Finally, optimal dose finding during drug development is not only able to streamline the drug development and improve drug product labeling, but also helps avoid postmarketing changes in labeling or even market withdrawals.

We express our gratitude to the Ernst Schering Research Foundation (ESRF) and Dr. Monika Lessl for their financial and logistic support of this workshop. We also wish to acknowledge the contributions of Dr. Bernd Müller, Head of Global Preclinical Development, Schering AG, as well Frau Yvonne Spiegel, Schering AG, and Frau Karola Szivos, ESRF in organizing the event. We gratefully recognize the invited presenters and moderators for their lectures, manuscripts, and discussion leads. Finally, we wish to thank the workshop audience for their insightful questions and comments that made this workshop a success.

Wolf Sittner
Jürgen Venitz

Contents

List of Editors and Contributors

Editors

Venitz, J.
Department of Pharmaceutics, School of Pharmacy,
Medical College of Virginia, Room 450B, R.B. Smith Building,
410 N 12th Street, P.O. Box 980533, VA 23298-0533 Richmond, USA
(e-mail: jvenitz@vcu.edu)

Sittner, W.
Clinical Pharmacology, Schering AG, Müllerstr. 178, 13342 Berlin, Germany
(e-mail: wolf.sittner@schering.de)

Contributors

Arjomand-Nahad, F.
Institute for Clinical Pharmacology, Charité – Universitätsmedizin Berlin,
Campus Charité Mitte, Schumannstr. 20/21, 10117 Berlin, Germany

Brockmöller, J.
Department for Clinical Pharmacology,
Universitätsklinikum der Georg-August Universität Göttingen,
Robert Koch Str. 40, 89081 Ulm, Germany

Buchan, P.
Nerviano Medical Sciences, Head of Clinical Development,
Viale pasteur, 10, 20014 Nerviano (Milano), Italy
(e-mail: peter.buchan@nervianoms.com)

Cascorbi, I.
Institute for Pharmacology, Universitätsklinikum Schleswig-Holstein,
Campus Kiel, Hospitalstr. 4, 24105 Kiel, Germany

Davis, J.
Clinical R&D, Pfizer Global Research and Development,
Sandwich Laboratories, Ramsgate Road, Sandwich, CT13 9NJ, UK

Gerloff, Th.
Institute Clinical Pharmacology, Charité – Universitätsmedizin Berlin,
Campus Charité Mitte, Schumannstr. 20/21, 10117 Berlin, Germany

Gimmi, C.
Global Pharma Development, Medical Sciences, PDM2,
F.-Hoffmann-La Roche Ltd, Grenzacher Str. 124, 4070 Basel, Switzerland
(e-mail: Claude.Gimmi@Roche.com)

Henze, G.
Charité Campus Virchow Klinikum,
Augstenburger Platz 1, 13353 Berlin, Germany
(e-mail: guenter.henze@charite.de)

Jackson, S.H.D.
Department of Clinical Gerontology, King's College Hospital,
Bessemer Road, London SE5 9PJ, UK
(e-mail: Stephen.Jackson@kcl.ac.uk)

Kirchheiner, J.
Abteilung Naturheilkunde und Klinische Pharmakologie,
Universitätsklinikum Ulm, Helmholzstr. 20, 89081 Ulm, Germany

Kuhlmann, J.
Bayer Health Care AG, Pharma Center,
Henselweg 22, 42115 Wuppertal, Germany
(e-mail: jochen.kuhlmann@bayerhealthcare.com)

Laschinski, G.
Institute for Clinical Pharmacology, Charité – Universitätsmedizin Berlin,
Campus Charité Mitte, Schumannstr. 20/21, 10117 Berlin, Germany

Leung, Th.
Comprehensive Oncology Centre, Hong Kong Sanatorium and Hospital,
2 Village Road Happy Valley, Hong Kong SAR, China
(e-mail: thomaswtleung@hksh.com)

Littman, B.H.
Clinical R&D, Pfizer Global Research and Development,
Sandwich Laboratories, Ramsgate Road, Sandwich, CT13 9NJ, UK

Machin, D.
UKCCSG-Data Centre, University of Leicester, Hearts of Oak Hous,
9 Princess Road West, Leicestr LE1 6TH, UK
(e-mail: david.machin1@onetel.net)

Marshall, S.
Clinical R&D, Pfizer Global Research and Development,
Sandwich Laboratories, Ramsgate Road, Sandwich, CT13 9NJ, UK

Morganroth, J.
University of Pennsylvania, School of Medicine and Chief Scientist,
eResearch Technology, Inc.,
30 South 17th Street, Philadelphia, PA 19103-4001, USA
(e-mail: Jmorganroth@ert.com)

Peck, C.
UCSF Center for Drug Development Science,
Room 211, U.C., Washington Center 1608,
Rhode Island Ave, N.W. DC 20036 Washington DC, USA
(e-mail: carl_peck@compuserve.com)

Price, P.
Academic Department of Radiation Oncology, Christie Hospital NHS Trust,
Wilmslow Road, Withington, Manchester M20 4BX, UK
(e-mail: pat.price@manchester.ac.uk)

Roots, I.
Institut für Klinische Pharmakologie, Charité – Universitätsmedizin Berlin,
Campus Charité Mitte, Schumannstr. 20/21, 10117 Berlin, Germany
(e-mail: ivar.roots@charite.de)

Rose, K.
F. Hoffmann-La Roche Ltd, Pharmaceuticals Division, PDM5,
4070 Basel, Switzerland
(e-mail: klaus.rose@roche.com)

Sarapa, N.
Daiichi Sankyo Pharma Development,
399 Thornall St., Edison, NJ 08837, USA
(e-mail: nsarapa@daiichisankyo-us.com)

Schwarz, D.
Institut für Klinische Pharmakologie, Charité – Universitätsmedizin Berlin,
Campus Charité Mitte, Schumannstr. 20/21, 10117 Berlin, Germany

Singer, Th.
Non Clinical Safety, Safety and Technical Sciences – STS,
Hoffman-La Roche Ltd. Grenzacherstrasse, 4070 Basel, Switzerland
(e-mail: Thomas.singer@roche.com)

Sultana, S.R.
Clinical R&D, Pfizer Global Research and Development,
Sandwich Laboratories, Ramsgate Road, Sandwich, CT13 9NJ, UK
(e-mail: Stefan.sultana@pfizer.com)

Tan, S-B.
Division of Clinical Trials and Epidemiological Sciences,
National Cancer Centre, Singapore 169610

1 Extrapolation of Preclinical Data into Clinical Reality – Translational Science

T. Singer

Abstract. Human and animal in vitro models are potentially powerful preclinical tools: prediction of pharmacological behaviour of drugs; selection of animal species most closely related to humans based on metabolic patterns; prediction of drug interactions and explanation of metabolic origins of interindividual variabilities in pharmacological activity. Extrapolation of preclinical data into clinical reality is a translational science and remains an ultimate challenge in drug development.

Preclinical in vivo and in vitro studies are fundamental to the safe and effective development of new drugs. Preclinical research is essential to a better understanding of the pharmacological and toxicological activities of drugs and their metabolites. Data generated by animal models and alternative methods can be used and extrapolated to improve clinical trials, particularly those for anticancer drugs.

Innovative drug therapy has led to an impressive drop in death rates for a variety of diseases in the past few years. In particular, outstanding results have been achieved in oncology. Data presented at the 41st ASCO Annual meeting[1] showed that the anticancer drug Bevacizumab (Avastin) led to longer survival and tumor control in patients suffering from colorectal cancer[2] and non-small cell lung cancer (NSCLC)[3]. A significant reduction in the risk of disease progression was demonstrated for patients with metastatic breast cancer[4]. Bevacizumab is a monoclonal antibody that works by attaching to and inhibiting the action of vascular endothelial growth factor (VEGF) in laboratory experiments.

Clinical trials with trastuzumab (Herceptin) – the first humanized antibody approved for the treatment of HER2-positive metastatic breast cancer – demonstrated a 50% reduction in the risk of disease recurrence. Trastuzumab is designed to target and block the function of HER2 protein overexpression. Research has shown that HER2-positive breast cancer is a more aggressive disease with a greater likelihood of recurrence, a poorer prognosis, and a decreased chance of survival compared with HER2-negative breast cancer.

However, besides the achievements of innovative drug therapy drawbacks also had to be faced. Various reasons such as adverse effects and drug interactions occurring after approval have led to a number of drug withdrawals in the past few years. Examples of such withdrawals that caused a high interest in both the lay and the medical press are the diet pill Fenfluramine/Dexafluramine (withdrawn in 1998 due to heart valve abnormalities), the antihypertensive drug Mibefradil (Posicor) (withdrawn in 1998 due to drug interactions and liver damage), the anticholesterol drug Cerivastatin (Baycol) (withdrawn for rhabdomyolysis in 2001), and recently the Cox-2-inhibitor Valdecoxib (Bextra), withdrawn due to cardiovascular reactions in 2005.

---

[1] ASCO 41st Annual Meeting of The American Society of Clinical Oncology, May 13–17, 2005, Orlando, Florida, USA

[2] 32% improvement in overall survival in patients with colorectal cancer (CRC) (E3200)

[3] 30% improvement in overall survival in patients with non-small cell lunger cancer (NSCLC) (E4599)

[4] 50% reduction in the risk of disease progression in patients with metastatic breast cancer (MBC) (E2100)

Challenges of the highly competitive environment in the pharmaceutical industry demand efficiency and optimized efforts in the key areas of science, quality, performance, and personnel management. One challenge of preclinical drug safety is to generate and to interpret information from different subdisciplines contributing to nonclinical safety testing, such as *in silico* tools, early metabolism, secondary pharmacology, experimental and mechanistic toxicology, mutagenicity, pharmacokinetics, and safety pharmacology.

Extrapolation to humans is achievable only when effects produced in appropriately qualified laboratory animals are relevant to humans. The exposure of experimental animals to high doses is necessary to discover possible hazards to humans. However, species differences, different physiology, metabolism, organotrophy (e.g., GI tract in dogs), and the setting of "healthy animal versus human patient" often make an extrapolation of data to humans very difficult.

Information generated by rapidly growing data bases, laboratory and screening technologies must be assessed in terms of risk vs hazard and risk vs benefit. With alternative methods becoming available for toxicological screening, a dramatic reduction in numbers of animals used in nonclinical studies could be achieved. Today, cytotoxicity screening can be done with cultures of retina, brain and meningeal cells, hepatocytes, and kidney cells.

Three examples for molecular methods applied in toxicology are dioxin, which tightly binds to a protein, the Ah receptor in mouse liver, which is a species-specific toxicity. Peroxisome proliferators cause liver tumors and testicular damage in rodents. Unleaded gasoline binding to α2-microglobulin induce rat kidney tumors. However, this was not considered a carcinogenic risk for humans by the EPA. Although molecular toxicology can deliver valuable results, it should not be overestimated. Molecular toxicology must not be performed for its own sake, if instead smarter results could be generated by animal data using the skills of traditional toxicology.

The main goals of toxicity evaluation are to produce *predictive and mechanistic outcomes*. Predictive outcomes are needed to create safe new drugs, to select the right drug candidates early, and to shorten overall development times by dropping compounds with toxic liabilities early. Hence, a reduction in costly late development phases and an improvement in success rates of developing safe and effective medicines can be achieved by concentrating on the development of medicines with the best quality. *Mechanistic outcomes* are needed to understand the reasons for observed side effects or toxicities and to accurately extrapolate from the preclinical to the clinical situation regarding safety issues.

Optimization of safe treatment regimes for different cancer drugs is extremely complex. The type and extent of a cancer will determine the method and schedule of administration of this drug. The combination of dose, route, cycle duration, and schedule is nearly unlimited. In addition, depending on the treatment cycle in a particular indication, the duration of drug holiday changes. Fluorouracil (5-FU) and Capecitabine (Xeloda), which are basically the same drug, have a completely different schedule. XELODA is a fluoropyrimidine carbamate with antineoplastic activity. 5-FU is one of the oldest chemotherapy drugs, and has been in use for decades. It is an active medicine against many cancers and is leading to leukopenic effects such as myelosuppression. The effects of an anticancer drug can be assumed to take place in the bone marrow based on irreversible linear or capacity-limited cytotoxicity.

The expected benefits of predicting myelosuppression in humans are the choice of the right starting dose with respect to safety and optimization for efficacy. This can be achieved using data generated with a generic model. The model can be applied across projects, may incorporate competitor data, and builds on previous experience and knowledge. We have developed a model describing the maturation of cells from the progenitor stem cells in the bone marrow to the circulating cells in the blood: the semi-mechanism-based pharmacokinetics/pharmacodynamics (PK/PD) model for myelosuppression. The focus is on myeloid cells, in particular to a fraction of granulocytes called neutrophils, since their depletion is the most common hematological adverse event seen with cytotoxics. The model helps predict the time course of myelosuppression in relation to the drug concentration vs the time profile. Proliferation and maturation stages of myeloid cells in the bone marrow and cell removal from the circulation can be quantified. Interestingly, the model works for both rat and human data. Furthermore, it is also possible to predict the myelosuppression after several cycles of administration, as shown for Paclitaxel (three cycles) and Vininflunine (two cycles).

Prediction of Myelosuppression from Preclinical to Human Data: A System Approach Data obtained from tolerability studies in rats can be used to estimate the potential of the drug for myelosuppression, i.e., the drug parameter. It is assumed that the drug parameter will not change across species. The difference in the physiology of the species is taken into account. A clear differentiation between the drug effects and the system effects (i.e., the physiology) must be made. Our assumption of the interspecies scaling is confirmed by data from the literature showing that a linear correlation between the drug parameter across species exists. We can then start from rat data, estimate the drug parameter, and then predict the drug effect in humans.

Conclusion Human and animal in vitro models are potentially powerful preclinical tools in the prediction of the pharmacological behavior of drugs; the selection of the animal species most closely related to humans on the basis of metabolic pattern; the assessment of the duration of drug action, particularly those drugs exhibiting different metabolic clearances; the understanding and prediction of drug interactions; and the explanation of the metabolic origins of interindividual variabilities in pharmacological activity. The ultimate challenge remains the extrapolation of preclinical data to clinical reality.

2 Smarter Candidate Selection – Utilizing Microdosing in Exploratory Clinical Studies

P. Buchan

Abstract. Microdosing offers a faster and potentially less expensive approach to obtaining human in vivo PK data in early clinical drug development. It encompasses the use of pharmacologically inactive doses of test drug in the low microgram range along with ultrasensitive assay methods (PET, AMS) to assess human exposure in order to extrapolate the PK of higher, clinically more relevant doses, assuming linear PK. This strategy allows early evaluation of systemic clearance, oral bioavailability as well as sources of intersubject variability and questions of specific metabolite formation. It does take advantage of reduced regulatory requirements of preclinical safety studies, bulk drug synthesis (CMC requirements) and easier formulation options, e.g., as part of an exploratory IND; however, this is counterbalanced by a need to synthesize radiolabeled test compound and the development of a sophisticated analytical method. Ongoing studies will determine the predictability of human PK using Microdosing methods.

2.1 The Need for Exploratory Clinical Studies in the Successful Development of New Drugs

As a result of advances in molecular biology to identify potential new drug targets and in combinatorial chemistry and high-throughput screening to synthesize and select ligands for these targets, the pharmaceutical industry now has the means to generate large numbers of drug candidates for clinical testing. However, of the compounds entering clinical studies it is estimated that still only about one out of 10–20, depending on the indication and class of compound, will progress to marketing approval. Frequently these candidate drugs reach a relatively late stage of development before being discarded, resulting in unsustainable losses of time and money. This late-stage attrition can kill projects, companies, jobs and, sadly, probably also patients.

Retrospective analysis has revealed that drugs fail for various reasons such as inadequate efficacy, unacceptable toxicity, or unmanageable pharmacokinetic (PK) properties or a combination of these. Feedback from the clinic has stimulated the development and application of more predictive nonclinical models to improve preclinical compound selection. For example, advances in our understanding of the PK clinical situation and the development of several predictive in vitro and now *in silico* models have allowed the elimination of molecules with generally undesirable properties such as potent inhibition of cytochrome P450 3A4. However, despite general improvements, particularly in the PK field, late-stage attrition still beleaguers the industry and PK issues are still a cause for concern. There must be other reasons for this situation.

One obvious problem is that there can be no universal standards for all drugs, and the desired properties in terms of efficacy, toxicity, PK convenience, etc. of each candidate drug will depend on various factors such as indication, patient population, and existing and future therapies. Thus an essential feature at an early stage of any project to develop a new drug is a clear definition of the standards in the form of a target candidate drug or product profile. Input is required from clinicians, scientists and business analysts to define a profile which, if achieved, will not only ensure marketing approval but also a return on investment once on the market, a feat achieved by only three out of ten registered drugs. As the properties of a candidate drug are investigated, the target profile serves as the point of reference for judging whether it is still capable of attaining the predefined profile for success or whether the chances are so low that it should be dropped from development. Obviously the earlier the key properties are studied, the earlier decisions can be taken on whether or not to continue development and thus reduce late-stage attrition. Nonclinical data may in some cases be sufficient to drop a candidate or give enough reassurance for investment in a full development program. However, despite significant improvements in nonclinical models, frequently they will not provide all the answers but instead draw attention to critical issues which may jeopardize the chances of a drug candidate meeting the target product profile, which must then be addressed with data obtained from study in humans.

Unfortunately, even with an awareness of these issues, adopting the traditional approach of drug development by predefined phases with corresponding studies still runs the risk of late identification and confrontation of issues and continued late-stage attrition. There is a clear need for an alternative approach that conceptually separates exploratory from full clinical development. Exploratory clinical development consists of early studies in humans to address key issues identified as critical to the development of a successful drug. It can be viewed as a risk management exercise, reducing the risk that a compound will be dropped at a late stage. This may not completely eliminate late-stage attrition but would improve the quality of compounds taken forward and allow entry into full development, with all its inherent costs and commitments, to proceed with a knowledge of the risks related to foreseeable issues. The Food and Drug Administration (FDA) in the United States expressed its awareness of this need and in its March 2004 Critical Path Report (Food and Drug Administration 2006), explained that to reduce the time and resources expended during early drug development on candidates that are unlikely to succeed, tools are needed to distinguish those candidates that hold promise from those that do not earlier in the process. The industry needs to act accordingly and adopt strategies to kill compounds earlier, i.e., kill or be killed.

Exploratory clinical studies would not always correspond to those in the full phase I package and may be abbreviated versions of investigations normally performed at a later stage or specifically designed innovative studies obviously respecting safety, ethical, and regulatory standards. Of the innovative approaches, human microdosing has received particular attention, as it offers a faster and potentially less expensive approach to obtaining human in vivo PK data (Lappin and Garner 2003; Wilding and Bell 2005; Sarapa 2003).

2.2 Concept of Microdosing

The fundamental concept of microdosing is that under conditions of PK dose proportionality, PK data obtained at low doses can be used to predict PK at higher doses. The application of this concept has been made possible by the development of extremely sensitive analytical techniques that permit the acquisition of pharmacokinetic or imaging data at low doses of candidate drugs unlikely to induce any tangible biological effect. Therefore the potential risk to human subjects is very limited and information adequate to support the initiation of such limited human studies can be derived from limited nonclinical safety studies. This opens the possibility for a nonclinical safety package reduced in terms of time and costs, permitting earlier in vivo investigation of candidate drugs with potential human PK issues.

Recognizing that the nonclinical safety guidance International Conference on Harmonisation (ICH) Topic M3 (ICH Harmonised Tripartite Guideline M3 2000) did not adequately cover this possibility, the European Medicines Agency (EMEA) produced a position paper in 2003 specifically addressing nonclinical studies required to support a microdose clinical trial (EMEA 2003). In the US, the FDA felt that sponsors had not taken sufficient advantage of the flexibility offered by existing guidances (such as that for single-dose acute toxicity testing for pharmaceuticals [FDA 1996]) when planning innovative early clinical trials and therefore issued a guidance in January, 2006, on Exploratory Investigational New Drug (IND) studies (FDA 2006), which also addresses human microdosing in a broader framework. The EMEA position paper proposed a definition of the term "microdose" that also appeared in the FDA guidance, namely,

> . . . the term "microdose" is defined as less than 1/100th of the dose calculated to yield a pharmacological effect of the test substance based on the primary pharmacodynamic data obtained in vitro and in vivo (typically doses in, or below, the low microgram range) and at a maximum dose of $\leq$ 100 microgram.

For reasons explained later in this chapter, I find this definition somewhat illogical and restrictive and I prefer a more flexible definition such as "a dose unlikely to induce any tangible biological effect" to which individual candidate drug characteristics can be applied to determine the most appropriate safety factor such as the nature of the biological effects, the ability to accurately measure these effects, and the likelihood of individual variability in responses.

The concept of microdosing applies not only to unlabeled test compounds but also allows for the administration of small amounts radiolabeled compound provided that this also respects an adequate safety factor. To derive meaningful data from such low doses of radioactivity, ultrasensitive techniques such as accelerated mass spectroscopy (AMS) and positron emission tomography (PET) must be applied. PET is discussed in detail elsewhere in this book and is therefore not addressed in this chapter.

In addition, a radioactive microdose may be "spiked" to a pharmacologically active dose of unlabeled drug and the safety of trial subjects respected if there is already adequate clinical experience with the drug.

2.3 Applicability and Advantages of Microdosing

Progress on three fronts, namely, analytical, regulatory and an understanding of the role of PK in drug development, has brought the industry to a position where a microdosing study can be considered as a possible first step in clinical investigations. The industry must now address the strategic issue of when and when not to conduct a microdose study. This is not simply a matter of evaluating when microdosing would be technically feasible and extrapolation to higher doses apparently valid, but of judging if the information gained would be valuable in deciding whether and how to advance a candidate drug through development. It is therefore useful to plan a decision path based on the possible outcomes of the study. Although this need not be binding in the final analysis, if a coherent decision path cannot be charted based on possible outcomes, then the decision to conduct the study or its design must be questioned. Microdosing is therefore applicable when it answers a need, when it is technically feasible, and when there is a reasonable chance of predictability. When there are no apparent PK issues, or when these can be adequately addressed in a timely manner by other approaches such as in vitro studies or standard single ascending dose human studies, a microdose study will not be warranted. By definition microdose studies provide information on the pharmacokinetics and metabolism of a compound but, by definition, not directly on its pharmacodynamics or toxicity.

Studies can be conducted on a single test substance or on a number of closely related compounds with the purpose of differentiating between these. The information can be gained on all PK parameters, including mass balance and metabolite profiling by AMS analysis and tissue distribution by PET analysis when radiotracers are used. It can thus be used to investigate questions of:

- Exposure in relation to dose and the associated absorption and clearance estimates
- Absolute bioavailability by the rapid provision of IV data by microdose (avoiding potential formulation and safety problems at higher IV doses)
- Intersubject variability of diverse origin
- Specific metabolite formation

Microdosing is obviously not appropriate for directly addressing issues resulting from nonlinearity at higher doses such as the saturation of presystemic metabolism or the capacity of formulations to resolve solubility problems.

The major advantage of microdosing is that it offers a quicker and less expensive means of obtaining key human PK data through a reduced requirement of preclinical safety studies, reduced bulk compound synthesis (and CMC requirements), easier formulation options, and less bureaucracy (e.g., using an exploratory IND).

However, because of the low drug concentrations associated with microdosing, there are bioanalytical aspects to consider before opting for this approach. The choice is either to push for lower limits of quantification when administering nonradiolabeled compound or to apply extremely sensitive radiotracer analytical methods, such as AMS, requiring synthesis of radiolabeled compound.

2.4 Prerequisites and Preparation for a Human Microdosing Study

The documents produced by the EU and US regulatory authorities that address human microdosing give guidance on prerequisites but are not binding nor comprehensive. They are referred to here as at present they provide the only general point of reference for sponsors of studies and ethical committees granting permission to conduct such studies. Some additional points not explicitly covered by these documents are addressed and further comments on their content are noted later in this section.

2.4.1 Nonclinical Studies to Support Single Microdose Studies in Humans

Both the EMEA and FDA documents make similar recommendations for an extended single-dose toxicity study in one species, the choice of which should be justified by comparative in vitro metabolism and primary pharmacodynamics (PD) data. The test compound should be administered by the intended clinical route, thus covering potential local tolerance problems; the EMEA also recommend IV administration when this is not the intended route.

In these studies, animals should be observed for 14 days after dosing with an interim sacrifice on day 2. Endpoints evaluated should include body weights, clinical signs, clinical chemistries and hematology on days 2 and 14, and histopathology. Gross necropsy should be performed on animals found dead, sacrificed moribund, and on days 2 and 14. The study should be defined with sufficient numbers, a control group, and consideration of the use of both genders in order to establish a dose inducing a minimally toxic effect.

However, in the case of compounds with low toxic potential, alternatives are proposed to avoid excessive dosing and use of animals. The EMEA proposes a limit dose approach for low toxicity compounds set by allometric scaling from animal species to man and using safety factor of 1,000, whereas the FDA proposes "establishing a margin of safety" by demonstrating that a large multiple (e.g., $100\times$) of the proposed human dose does not induce adverse effects in the experimental animals and that scaling based on body surface area can be used to select the dose for use in the clinical trial.

Although a full safety pharmacology package is not requested by either authority, the EMEA requests the provision of all other available background information relevant to vital organ function and safety parameters on the test substance during screening and/or on closely related drugs (e.g., receptor screening profiles, hERG, APD, and behavioral screens).

Whereas the EMEA recommends conducting genotoxicity studies according to ICH guidance allowing for abridged or reduced testing for well-known chemical classes if scientifically justified and providing valid data, the FDA position is more relaxed, stating, "as exposures are comparable to routine environmental exposures, routine genetic toxicology testing is not needed."

Logically the authorities note that any observed toxicity may require further clarification and that all nonclinical studies should be conducted according good laboratory practices (GLP).

2.4.2 Test Article

The FDA is in the process of developing guidance explaining the stepwise approach to meeting current good manufacturing practice (CGMP) regulations, which will specifically address the manufacture or preparation of products intended for use in an exploratory IND study. In the meantime, applicable Chemistry, Manufacturing and Controls (CMC) information is discussed in the draft guidance. The intention is to make fewer demands than those made for traditional INDs, for example, less impurity characterization is needed if the same batch of candidate product is used in both the toxicology studies and clinical trials. The EMEA position paper does not address CMC for the test article and currently it unclear how different member states will approach microdosing in the light of the new Clinical Trials Directive.

2.4.3 Bioanalytical Methodology

Because of the low drug concentrations associated with microdosing, highly sensitive bioanalytical methods are required for the accurate determination of PK parameters. It is therefore prudent to estimate the likely critical concentrations and ensure an adequate lower limit of quantification for accurate measurement. As the results are for internal decision making and will not constitute a critical element in a regulatory dossier, a validated analytical method is not necessary but rather a robust, fit-for-purpose method. Concentration estimates can be made from a PK model predicted from available preclinical in vitro and in vivo data, the intended dose and sampling times. For example, if a simple one-compartment model, with first-order input and output, absorption and elimination half-lives of 0.5 h and 12 h, respectively, volume of distribution of 1 l/kg is predicted and a 70-kg person receives a 100-μg oral dose that is 100% absorbed, the maximum plasma concentration will be about 1 ng/ml and the concentration at 24 h after the dose, about 400 pg/ml. Although in most cases this does not present an analytical challenge with today's methods, the concentrations to be measured could be considerably lower if a lower dose is selected for safety reasons, absorption is less than 100%, the volume of distribution greater than 1 l/kg, and elimination occurs at a faster rate or is polyexponential.

In addition, given the vagaries of extrapolating human models from preclinical data and the variability between human subjects, it is advisable to set the target lower limit of quantification (LLOQ) two- to threefold below the lowest estimated concentration to be measured for obtaining reasonable PK estimates.

When there is a risk that adequate estimates of PK parameters will not be obtained because target LLOQ cannot be achieved using nonradiolabeled analytical methods, AMS now offers an alternative solution. The extreme sensitivity AMS lies in the fact that it separates and counts individual atoms whereas liquid scintillation counting (LSC) measures decay events. With a $t_{1/2}$ of 5,730 years, about 4×10^9 atoms of ^{14}C are required to produce one disintegration per second. A five-million-volt Van der Graaff tandem accelerator mass spectrometer requires the detection only about 1,000 ^{14}C atoms to determine the isotope ratio of $^{12}C/^{14}C$ for a precise count permitting measurement in the attomole to zeptomole (10^{-18}–10^{-21} mole) range. As a practical example, in a profiling study, a metabolite peak equal to 0.006 disintegrations per minute could be easily detected. Human studies relying on LSC tend to use radioactive doses near ethical limits often around 80–100 μCi/subject determined by dosimetry calculations using data from animal tissue distribution studies. As the typical dose for an AMS study is 100 nCi, similar to the natural background level, this is considered as safety and requires no prior tissue distribution/dosimetry studies.

Although AMS is in a league of its own in terms analytical sensitivity, it does require synthesis of a radiolabeled version of the test compound, which can in some cases be problematic. This may have time and cost implications. Analytical costs using AMS are also high because of the highly sophisticated equipment involved and the need for sample fractionation to measure parent compound and individual metabolites.

However, microdosing with AMS offers a unique way of resolving human PK issues at an early stage (with potential savings greatly in excess of analytical costs!) and is appropriate not only when nonradiolabeled methods do not meet sensitivity requirements, but also for addressing metabolite profiling issues, a problem that can be solved by total radioactivity measurements (metabolically stable molecules, routes of elimination) and specific situations such as analysis of tissue and fecal samples.

2.4.4 Pharmacokinetic Prerequisites

Although the existing regulatory documents give little in the way of specific instructions on nonclinical PK, certain prerequisites can be deduced by inference from the EMEA document. Comparative in vitro metabolism can be performed to provide data in support of the species chosen for the single-dose toxicity study. Allometric scaling from animal species to humans is recommended when a limit dose approach is adopted, implying that PK/toxicokinetic studies, including plasma protein binding, should be conducted in more than one animal species. The CPMP statement that "[toxicity] information should also be obtained on any other organ system where the test substance localises and e.g., those organ systems intended to be visualised by imaging agents" is unclear and could be interpreted as a need for tissue distribution studies to identify organ systems where the test substance localizes. This seems excessive and is in contradiction with the aims of expediency and reduced animal use.

Aside from the regulatory documents, as the decision to conduct a microdose study requires a solid rationale, sufficient PK/metabolism data must be generated to define the need for, objectives, and design of the human microdose study. These data may come from routine studies included in early PK/metabolism screening programs and from specifically designed studies based on the results of the screening program or on the intended use of the candidate drug. Examples of such studies are those that predict dose–exposure relationships from bioavailability and clearance estimates or predict routes of elimination, such as specific enzyme pathways.

2.5 Future of Microdosing

The extent to which microdosing will be used in the future will depend on several factors:

- Confidence in its predictability of PK at higher (therapeutic) doses and conversely its limits
- The time and costs involved – on the one hand in conducting nonclinical safety studies and on the other in developing suitable bioanalytical strategies
- Awareness of the need to identify and address key PK issues as early as possible and of the scope of its application

2.5.1 Determining Predictability

Data derived from a microdose study are only of value when they can be used to predict PK parameters of interest at higher anticipated therapeutic doses. Thus dose-exposure proportionality should apply when information on dose-related PK parameters, such as C_{max}, AUC, and clearance (dose/AUC), is required. However, even in situations of dose-dependent kinetics such as nonlinear absorption, data derived at microdoses may still be useful to determine rate constant-type parameters (e.g., $t_{1/2}$), providing these do not vary with dose and exposure.

Dose–exposure proportionality is normally seen when PK follows bulk processes dependent on physicochemical properties of a soluble drug.

The vast majority of nonlinear PK situations have been observed at the upper end of the dose range of drugs in association with saturation of rate-limiting processes such as dissolution of solid drug forms or metabolism by specific enzymes with limited capacity, either presystemically during the absorption phase or during the elimination phase. These types of nonlinearity are frequently observed at high doses used in nonclinical toxicology studies but have also been observed within the therapeutic range of certain drugs in humans. It is obvious that in such cases data from microdose studies can only be used to predict exposure within the range of doses where linear PK is observed but not beyond into situations of nonlinearity.

However, when such issues are expected, human microdose studies may still be useful in answering specific questions rapidly. For example, when solubility of a drug candidate is expected to pose a problem at higher doses and clearance in humans cannot be readily predicted from nonclinical data, a microdose could be used to measure clearance and thus estimate the amount of drug that must be absorbed to achieve the predicted therapeutic exposure. From the solubility data, the feasibility of formulating the drug to attain the estimated required absorption could be evaluated and hence decisions could also be made on whether or not to proceed with development. Similarly, microdosing may also fulfill a role when nonclinical data indicate that a drug may be subject to Michaelis-Menton kinetics through metabolism by specific enzymes and thus lead to nonlinearity at higher doses but cannot indicate with certainty the relative importance of the enzyme in the overall elimination of the drug in humans. In this case, a microdose study in phenotyped and genotyped subjects selected for their differing levels of activity of the enzyme of interest could provide more direct information on the relative importance of this enzyme in the overall elimination and thus the likelihood that nonlinear PK will occur within the therapeutic dose range.

In contrast, very little is known about nonlinearity at the lower doses of drugs. Indeed, this is not surprising, as in the past low doses have been of little interest since they had no measurable biological activity and required extremely sensitive analytical techniques to study PK. The concern is whether, at very low doses, processes normally masked by bulk processes at higher doses can have a relatively greater influence on parameter estimates and thus exhibit nonlinearity. If a phenomenon of lower dose nonlinearity exits it is important for application of microdosing to know at what dose levels it might occur. Transporter mechanisms and binding to high affinity, low-capacity macromolecules have been speculated as processes that could lead to nonlinearity at very low doses. To reduce speculation, retrospective and more recently prospective analyses have been carried out.

To explore the utility of microdosing at Pfizer, a retrospective analysis of in-house phase I data has been done (Smith et al. 2003). The PK profile of 12 (recently evaluated) compounds at a low dose (usually < 5 mg) was compared to that at pharmacologically active doses. In six cases, more or less linear PK was observed and therefore microdosing would have probably been predictive. However, it was noted that of these six, two would have had predictable human PK from preclinical data and the other four were active at low doses – which would have posed (nonradiolabeled) analytical problems at a safe microdose level. The other six displayed nonlinear PK and microdosing would have given misleading data. Such retrospective analyses have merit and it would be of interest to view a larger range of similar data both from Pfizer and other companies. This approach would be even more enlightening if presented with more relevant details on the drug (for example, physicochemical, enzyme, and transporter substrate properties), routes of administration, and the reliability of the bioanalytical methods at lower doses.

In 2004–2005, the question of dose–exposure proportionality was addressed prospectively in the CREAM Trial (Consortium for Resourcing and Evaluating AMS Microdosing) instigated by the AMS CRO, Xceleron Ltd and the early clinical phase CRO, Pharm Bio-Research BV, with the participation of Eli Lilly, F Hoffmann la Roche, Schering and Servier. Professor Malcolm Rowland chaired the Scientific Advisory Group.

Table 1. Summary of CREAM trial

Test compound	Selection characteristics	Microdose result
Midazolam	A selective substrate for CYP3A4. High first pass metabolism	Predictive: excellent correlation of key PK parameters
Diazepam	Basic compound, substrate for CYP2C19. High plasma protein binding, extensive distribution and low clearance. Linear PK over a range of doses but possibly not at microdose	Predictive: excellent correlation of key PK parameters
ZK 253 (Schering drug candidate dropped after phase I)	Low bioavailability ($<1\%$) found in humans was difficult to predict due to variability in animal and in vitro data	Predictive: clearance, high volume of distribution and extremely low bioavailability was also predicted by microdosing
Warfarin	A CYP2C9 substrate. Stable in vitro but is extensively although slowly metabolized in humans	Not predictive, though slow metabolism and long half-life identified
Erythromycin	Substrate for both CYP3A4 and the intestinal efflux transporter P-glycoprotein	Declared "no test" as the oral microdose was degraded in the stomach due to acid lability

The aim of the trial was to assess the predictability of PK at therapeutic doses from human microdosing using compounds expected to strongly challenge the microdosing concept. Five test compounds were selected with complex metabolism-clearance properties for which animal or in vitro models did not satisfactorily predict human PK at therapeutic doses. Each compound was administered at a microdose level and at a therapeutic dose level, orally or intravenously as appropriate, to subjects in a cross-over design. Plasma concentrations were measured by AMS for all microdoses and the therapeutic doses of warfarin and diazepam. Therapeutic doses of midazolam, ZK253, and erythromycin were measured by LC-MS. The results of this trial have been presented at recent conferences (details are available at www.xceleron.com) and will be submitted to *Clinical Pharmacology and Therapeutics* for publication. The main findings are summarized in Table 1 and the CREAM sponsors have presented the following preliminary conclusions:

For three of the five selected drugs, human PK data obtained at a microdose were highly predictive of the situation at the corresponding therapeutic dose. The lack of dose–exposure linearity with warfarin was attributed to a high-affinity, low-capacity binding site and a low volume of distribution. However, the low clearance of warfarin was predictable from microdose data. Two of the drugs deviated from linear pharmacokinetic behavior, but nevertheless the microdose results gave useful insights into the properties of the drugs. The experience with erythromycin illustrates a potential limitation of microdosing but does not necessarily exclude its application to such compounds, provided potential problems are identified and precautionary measures are taken ensure that useful data can be obtained. For example, in this case, intravenous administration or the use of an enteric coated formulation or absorption site targeted delivery for the microdose may have produced data on certain PK parameters excluding, of course, those on dose–exposure relationships of a standard oral formulation.

Given that particularly difficult compounds were selected for the CREAM trial, it can be concluded from the three clearly positive outcomes that microdosing has promise as a predictive tool. Further publication of similar comparative data on other compounds will be of great help in the intelligent application of this predictive tool.

2.5.2 Regulatory Position

Without doubt, the publication of official documents by the US and European regulatory authorities making specific reference to a human microdose has helped to increase general awareness and lent credence to the concept and application of microdosing. This has provided a useful point of reference for ethical committees and for pharmacokinetists and other scientists trying to present a case within their companies for performing microdosing studies. However, sponsors generally tend to apply the recommendations made in guidances as they stand, without seeking to challenge the recommendations or adapt to particular situations despite, at least in the case of the FDA, an insistence on the nonbinding nature of guidances and an invitation to sponsors to argue for specific cases. Despite incurring extra costs and time in preparatory work this "belt and braces" approach is seen by sponsors as improving the likelihood of rapid approval by internal and external bodies including ethical committees and also the FDA. It is worth recalling that microdose studies are conducted to help make internal decisions and are unlikely to form a critical part of a dossier submission being superseded later in development by data obtained at more clinically relevant doses. Although sponsors could conduct preliminary work and design microdosing studies according to each individual case, the wording of guidance texts has a powerful impact. Therefore the regulatory authorities should seek to promote the intelligent use of such novel approaches by further improving initial texts based on further debate and experience with a new drug candidates. There are a number of points in these documents that already merit further reflection:

The following statement discussed earlier in this chapter is critical and open to criticism:

> . . . the term "microdose" is defined as less than 1/100th of the dose calculated to yield a pharmacological effect of the test substance based on the primary pharmacodynamic data obtained in vitro and in vivo (typically doses in, or below, the low microgram range) and at a maximum dose of $\leq$ 100 microgram.

Firstly, the phrase ". . . less than 1/100th of the dose calculated to yield a pharmacological effect" needs expansion and clarification. It is normal practice to predict the systemic exposure from preclinical data, and hence the dose required to achieve a therapeutic efficacy in humans. Depending on the mechanism of action, the therapeutic efficacy may be achieved at different degrees of pharmacological activity, and the steepness of the dose or concentration–response curve may vary. Thus the phrase "a pharmacological effect" may be interpreted as the pharmacological effect related to therapeutic efficacy rather than "a minimum pharmacological effect." Moreover, other undesirable effects may occur at lower doses than those required for therapeutic efficacy, as is the case for many anti-cancer drugs. The design of the nonclinical single-dose toxicity study is discussed in some detail in both the EMEA and FDA documents, including the need to establish "the dose inducing a minimal toxic effect," but there is no explanation of how to relate this dose to the calculation of the acceptable human microdose nor on the use toxicokinetic data to relate exposure to effect. A more comprehensive statement would be ". . . < 1/100th of the dose calculated to yield a biological effect" where "biological effect" refers to any effects related to primary pharmacological, secondary (including safety) pharmacological activity and toxicity. A topic for future debate when more experience has been gained is the adaptation of the arbitrary safety factor of 100 to the nature of the biological effects, e.g., innocuous or serious and the sensitivity with which these can be measured.

Secondly, fixing the dose ceiling arbitrarily at 100 μg for safety reasons is not logical because it does not take into account:

- The potency of the compound (e.g., for a compound with a pharmacological activity at a dose of 1,000 mg, a microdose of 100 μg would have an excessive safety factor of 10,000)
- The molecular weight (MW) (100 μg of a compound with a MW of 200 will contain 100 times more molecules than a compound of a MW of 20,000)
- The body weight of the subjects

The consequence of restricting the upper limit to 100 μg is that if combined with awkward PK such as incomplete absorption, large volume of distribution, or rapid elimination, the resulting plasma concentrations will be so low as to strain all current analytical techniques with the exception of AMS. In turn, AMS may not always provide the ideal solution if there are problems of radiosynthesis and separation of parent drug from metabolite, with the associated time and cost considerations. Adherence to a maximum dose of 100 μg may dissuade sponsors from conducting microdose studies that could have otherwise been conducted safely at higher doses.

Standardizing the EMEA and FDA documents in the way they address the need for safety pharmacology, genotoxicity, and a second route of administration in the single-dose toxicity study would also make life easier for sponsors intending to conduct microdose studies.

Both regulatory documents allow for only a single exposure in a human microdose study. However, there are situations where a microdose study could take advantage of a within-subject cross-over design, for example to study diurnal variation or absolute bioavailability when only one analytical method is available. It would be relatively easy to extend guidances to cover this possibility by allowing a second microdose after complete washout or at least permitting half of the maximum allowable dose to be administered twice.

2.5.3 Analytical Advances

Great strides have been made in the sensitivity of nonradiolabeled analytical methods and techniques such as LC-MS and MS, which are capable of responding to most requirements of a standard drug development program. However, the upper dose limits for microdosing proposed by the regulatory authorities has put a strain even on these techniques for many candidate drugs and, even when possible, the time required to adapt the methods may preclude the option of a microdose study in a development plan. All further advances in bioanalytical sensitivity will therefore increase the feasibility of conducting microdose studies. These may come from existing analytical techniques such as improvements in sample background clean-up or from new techniques such as molecularly imprinted polymers (Andersson 2000).

AMS can provide the answer to analytical sensitivity, but associated costs could daunt prospective users. Cost reductions may be achieved with the introduction of less powerful custom-built instruments dedicated to ^{14}C analysis (e.g., 500-kV accelerated mass spectrometers) and intelligent fraction sample selection for both parent compound analysis metabolite profiling.

2.5.4 Future Applications

To date the primary application of microdosing has been to gain early human PK data mostly in healthy male adult volunteers.

Given that a microdose is selected to be devoid of biological activity, it could provide a safe alternative to obtain PK data in populations considered as potentially vulnerable to toxic effects such as pediatric subjects and women of child-bearing potential. It may also be possible to give cocktail metabolizing enzyme probe substrates in microdose amounts to investigate genotypic and phenotypic variation or a drug's potential to inhibit these enzymes. If substrates were radiolabeled, AMS could serve as a single analytical method, following the requisite analyte separation, of course. The use of microdosing probes would have to be validated, for example in cross-over studies against current standard probe doses.

Microdosing may be particularly useful in overcoming some of the current difficulties in the development of anti-cancer drugs. Because of the potential toxicity of many anti-cancer drugs, typical first-in-human (FIH) studies employ an ascending dose schedule in small cohorts of cancer patients with the aim of determining the dose-limiting toxicity and maximum tolerated dose. Also for safety reasons, a low starting dose and relatively long observation periods before conservative dose escalation are employed with the result that few patients receive a therapeutically useful dose and much time is required to accumulate data. Consequently, human PK data are acquired slowly and randomly, i.e., dependent on patient eligibility and willingness to participate in clinical trials, such that specific PK problems are difficult to identify and address in a timely manner.

The safety margin provided by microdosing offers the possibility of conducting FIH studies with anti-cancer agents in healthy volunteers. The reliable and rapid recruitment of these subjects at specialized centers offers far more rapid generation of data to address issues related to PK and metabolism. In addition, data on the relationship between dose and exposure could be useful in designing of subsequent phase I clinical studies, e.g., in the selection of the starting dose and the escalation plan, thus improving the efficiency of the study in terms of the proportion of patients receiving an efficacious dose and of the time required to reach the study's endpoints. Healthy volunteers could also be taken from selected populations to investigate variations, such as those related to specific genotypes and phenotypes or renal and hepatic impairment.

AMS offers a further possibility for investigating PK in cancer and other patient groups undergoing therapy. The therapeutic doses could be spiked with very small amounts of radioactive compound as a specific study or as part of another study. These small radioactive amounts required for analysis by AMS present no special safety or contamination problems. This could constitute a convenient way of generating PK data, including mass balance and metabolite profiling, in a target patient population.

2.6 Conclusion

With the continued failure of drugs at late stages of development, there is a need to change from the current approach of development in predefined phases with corresponding studies and designs to an approach of identifying key issues that could undermine the progress of a candidate drug and address these as early as possible. For issues that can only be addressed by human in vivo studies, there is a need for creative thinking and the application of novel technologies and study designs in early exploratory clinical investigation. By virtue of the low dose and exposure of microdosing, fewer preclinical safety data are required, permitting earlier in vivo investigation of candidate drugs with potential human PK issues with negligible safety risks. Although the primary function of microdosing can be regarded as a risk management tool to select compounds with a lower chance of failure in later clinical development, it offers other benefits such as early feedback on preclinical models and reduction of the use of animals. It may provide early data to help in the design of subsequent studies, for example in selecting the starting dose for ascending-dose studies, thus reducing the number of dose escalations, and in planning more effective dosage regimens. This will result not only in fewer subjects being exposed to subefficacious doses but also in potentially reducing development times, thus allowing patients quicker access to safe, more efficacious drugs.

The intelligent and successful application of human microdosing will depend on changes in traditional thinking and on the collaboration between the sponsor in the pharmaceutical industry, service providers, and regulatory authorities.

References

Andersson LI (2000) Molecular imprinting: developments and applications in the analytical chemistry field. J Chromatogr B Biomed Sci Appl 745:3–13

EMEA (2003) Position paper on non-clinical safety studies to support clinical trials with a single microdose CPMP/SWP/2599/02, 23 Jan 2003. http://www.emea.eu.int/pdfs/human/swp/259902en.pdf. Revision 1, 23 June 2004, cited 6 April 2006

FDA (1996) Guidance for Industry: Single Dose Acute Toxicity Testing for Pharmaceuticals. CDER, August 1996. http://www.fda.gov/cder/guidance/index.htm

FDA (2005) Exploratory IND studies (contains nonbinding recommendations draft – not for implementation) April 2005 Pharmacology/Toxicology. http://www.fda.gov/cder/guidance/index.htm. Cited 6 April 2006

Food and Drug Administration (2006) FDA Critical Path Initiative – The critical path to new medical products. www.fda.gov/oc/initiatives/criticalpath/. Cited 6 April 2006

ICH Harmonized Tripartite Guideline (M3) (2000) Nonclinical safety studies for the conduct of human clinical trials for pharmaceuticals, issued 16 July 1997 and amended, 09 November 2000. http://www.ich.org/LOB/media/MEDIA506.pdf. Cited 6 April 2006

Lappin G, Garner RC (2003) Big physics, small doses: the use of AMS and PET in human microdosing of development drugs. NatRev Drug Discov 2:233–240

Sarapa N (2003) Early human microdosing to reduce attrition in clinical drug development. American Pharmaceutical Outsourcing Sept/Oct:1–5

Smith DA, Johnson DE, Park BK (2003) Use of microdosing to probe pharmacokinetics in humans – Is it too much for too little? Curr Opin Drug Discov Devel 6:39–40

Wilding IR, Bell GA (2005) Improved early clinical development through human microdosing studies. Drug Devel Today 10:890–894

3 The Applications of Biomarkers in Early Clinical Drug Development to Improve Decision-Making Processes

J. Kuhlmann

Abstract. Selecting and evaluating biomarkers in drug discovery and early drug development can substantially shorten clinical development time or the time to reach a critical decision point in exploratory drug development. Critical decisions such as candidate selection, early proof of concept/principle, dose ranging, development risks, and patient stratification are based on the appropriate measurements of biomarkers that are biologically and/or clinically validated. The use of biomarkers helps to streamline clinical development by determining whether the drug is reaching and affecting the molecular target in humans, delivering findings that are comparable to preclinical data, and by providing a measurable endpoint that predicts desired or undesired clinical effects and will increase the success rate in the confirmatory stage of clinical development. Appropriateness of biomarkers depends on the stage of development, development strategy, and the nature of the medical indication. Even if a biomarker fails in the validation process there may be still a benefit of having used it. More knowledge about pathophysiology of the disease and the drug has been obtained. Different levels of validation exist at different development phases. Biomarkers are perhaps most useful in the early phase of clinical development when measurement of clinical endpoints may be too time-consuming or cumbersome to provide timely proof of concept or dose-ranging information. Examples of biomarkers are illustrated for the development of new drugs in variant cardiovascular, pulmonary, and CNS diseases.

3.1 Introduction

Using new technologies in drug discovery such as genomics, high-throughput screening, and combinatorial chemistry, drug companies will see an explosion in the number of targets and leads it can explore (Kuhlmann 1997). Because of the continuous increase in time and cost of drug development and the considerable amount of resources required, pharmaceutical companies can no longer afford to continue to the resource and cost-intensive late phase of clinical trials with drugs that are unlikely to be therapeutically effective or that are not superior to existing treatments (Kuhlmann 1999; The CMR International R&D 2004).

Therefore, a tough selection process for picking candidate compounds out of research and a quick killing process for the candidate drug that does not measure up in advanced trials is more than ever mandatory to reduce attrition rates in late phases of development and to avoid wasting time, energy, and money. Or, in other words, the pharmaceutical industry must improve the quality of the drugs it produces by being much more selective at a much earlier stage in the research and development process to ensure that only optimal candidates get the benefit of full development resources (Haak van den et al. 2004). Strategies to improve the quality of decisions in drug discovery and development are the use of new tools and technologies such as pharmacogenomics to improve our knowledge on the origin of the disease and to identify new therapeutic strategies, modeling, and simulation of preclinical and clinical trials to bridge the gap between the early stages of the development of a new drug and its potential effects in humans, and selecting and evaluating biomarkers or surrogate markers for efficacy and safety (Kuhlmann 1999; Lesko et al. 2000; Danhof et al. 2005). Biomarkers will allow more integrated application of discovery and development information, leading to better clinical study design.

3.2 Definition and Classification

The NIH Biomarkers Definitions Working Group (Atkinson et al. 2001) has defined a biomarker as a characteristic that is objectively measured and evaluated as an indicator of normal biological processes, pathogenic processes, or pharmacologic responses to a therapeutic intervention. A surrogate endpoint is defined as a biomarker intended to substitute for a clinical endpoint, and a clinical endpoint is a characteristic or variable that reflects how a patient feels, functions, or survives.

A biomarker of drug effects should reflect a process on the critical path between the pharmacological action of the drug and its effect on a disease. Biomarkers may be divided into three distinctive groups: pharmacological, which can be observed in healthy volunteers; toxicological, which can also be observed in patients; and pathological, which can only be observed in patients having the disease (Danhof et al. 2005). This has important implications for their evaluation.

Another approach to the classification of biomarkers is on the basis of "directness" (Table 1) and a third approach is based on the notion that the overall mechanism of action of a drug depends on a series of events (Table 2). The last two approaches bear a large degree of similarity. On the basis of this, a new classification of seven types of biomarkers has been proposed by the COST B15 working group 2 "Markers of pharmacological and toxicological action" based on the location of the biomarker in the chain of events from underlying subject genotype or phenotype through to clinical scales (Table 3; Danhof et al. 2005).

According the guideline for industry "Exposure response relationships (2003)" biomarkers differ in their closeness to the intended therapeutic response or clinical benefit endpoints, including the following:

Table 1. Classification of biomarkers (based on "directness")

Biochemical or other measures of drug action at the molecular target level
Physiological measures
Pathological assessments
Clinical ratings

Table 2. Classification of biomarkers (based on mechanism of action)

Binding to the drug target (level 1)
Activation of the target (level 2)
Clinical response (level 3)

Table 3. New classification of biomarkers (Danhof et al. 2005)

Type 0	Genotype/phenotype
Type 1	Drug/metabolite concentrations
Type 2	Molecular target occupancy
Type 3	Molecular target activation
Type 4	Physiological measures
Type 5	Pathophysiological measures
Type 6	Clinical ratings

- Biomarkers thought to be valid surrogates for clinical benefit (e.g., blood pressure, cholesterol, viral load)
- Biomarkers thought to reflect the pathologic process and be at least candidate surrogates (e.g., brain appearance in Alzheimer's disease, brain infarct size, various radiographic and isotopic function tests)
- Biomarkers reflecting drug action but of uncertain relation to clinical outcome (e.g., inhibition of ADP-dependent platelet aggregation, ACE inhibition)
- Biomarkers that are still more remote from the clinical benefit endpoint (e.g., degree of binding to a receptor or inhibition of an agonist).

3.3 Why Do We Need Biomarkers?

Selecting and evaluating biomarkers in drug discovery and early drug development can substantially shorten development time or the time to reach a critical decision point in exploratory drug development. Even if a biomarker fails in the validation process there may be still a benefit of having used it. More knowledge about pathophysiology of the disease and the drug has been obtained. Advantages of using biomarkers in drug discovery and preclinical development including toxicology are summarized in Table 4.

Increasing use of biomarkers in early clinical development helps to streamline development by relating clinical data to preclinical test systems, determining whether the drug is reaching and affecting its molecular target in a human patient, and modeling phase I and phase II/III clinical trials. Further advantages of using biomarkers are:

- Improvement of the quality of decision making
- Reduction of the uncertainty
- Support of innovative products where previous knowledge and experience is insubstantial
- Selection of the best candidate
- increased success in late phase of clinical trials (confirmatory phase)
- Reduction of costs

Using biomarkers in the confirmatory stage of drug development helps to establish the dose–concentration–response relationship in patients with the target disease, to monitor efficacy and safety, and to identify subpopulations and new or modified indications. It should be mentioned, on the other hand, that biomarkers cannot guarantee success in the confirmatory stage of drug development, are not available for all indications, and have no direct benefit to healthy volunteers and patients. The ideal biomarker

Table 4. Advantages of using biomarkers in drug discovery and preclinical development

Identification of targets
Optimization of candidates (efficacy and safety)
Hints for biomarkers in humans
Pharmacological activity identification
Enhanced hazard identification
Species-bridging biomarkers
Biomarkers for simultaneous testing of drug efficacy and toxicity in animal models or in in vitro systems
Use of transgenic animal models for noninvasive imaging of target organ toxicity
PK/PD modeling (dose selection and escalation)

should be a specific, reproducible, sensitive, noninvasive measure of a disease and should be well understood and rapidly responsive to drug-induced changes. Normal values should be available and the variability should be preferably small and well-known. Biomarkers are not a substitute for measuring the clinical benefits of therapeutic interventions in the confirmatory stage of drug development. Appropriateness of biomarkers depends on the stage of discovery and development, the discovery strategy, and the nature of the medical indication.

3.4 Validation

One key obstacle relates to the issue of validation. Unfortunately, the validation term covers both the numerical precision aspects of an assay or technique and the confidence in the predictions one can make based on the data. Validation of biomarkers has to consider both sources of error and variability within the measurement itself as well as the nature of the biomarker in relation to the exposure–response continuum for the effect of interest or concern. The numerical precision aspects are well covered by standard approaches that come from guidelines on laboratory techniques (Workshop on Bioanalytical Methods Validation for Macro-molecules: Workshop Report 2001). These include reliability, sensitivity, specificity, relevance, accuracy, temporality, practicability, applicability, and observer bias. However, it is unlikely that a similar rigorous approach will be applicable to the ability to make predictions. Such predictions will be time-dependent as new information and data are obtained during the drug-development process (Rolan 1997). Biomarkers used as part of internal company decision making in early development require validation to a level consistent with their intended application. The criteria for validation are defined by:

- The nature of the question that the biomarker is intended to address
- The degree of certainty that is required for the answer
- The assumptions about the relationship between changes in the biomarker and clinical endpoints

The extent of validation should be determined by the strength and nature of the claims. Different levels of validation exist at different development phases. Biomarkers should be identified or generated during the discovery phase. It should be made a mandatory deliverable of the lead optimization stage. They should be prevalidated during the preclinical and early clinical phases regarding assay specificity, sensitivity, precision, accuracy, etc. During the late clinical phases, the principles of good laboratory practice (GLP) requirements are useful standards for validation of biomarkers (Colburn 2000).

3.5 Regulatory Aspects

The FDA has a legal basis for using surrogate endpoints in ordinary and accelerated drug approvals leading to market access of new drugs or drug products (US Government Printing Office 1997 Code of Federal Regulations; US Government Printing Office 1997 Federal Food, Drug and Cosmetic Act). The standards for linking a biomarker to a clinical outcome are higher for ordinary approvals than for accelerated approvals. This difference is based on consideration of many factors, including:

- The degree of scientific evidence needed to support biomarker surrogacy
- Public health needs
- Relative risk–benefit ratio
- Availability of alternative treatments (Lesko and Atkinson 2001)

For ordinary approvals, there are relatively few well-established surrogate endpoints; the FDA has approved drugs that:

- Lower cholesterol and triglycerides (coronary artery diseases)
- Lower arterial blood pressure (stroke, heart attacks, heart failure)
- Change bone density (osteoporosis)
- Increase cardiac output (acute heart failure)
- Reduce ventricular premature beats (for use in symptomatic patients)
- Reduce HIV-RNA load and enhance $CD4^+$ cells (AIDS)
- Lower blood sugar and glycohemoglobin (diabetes mellitus)
- Lower plasma testosterone levels (prostate Ca)
- Reduce tumor size (cancer)

Regarding the FDA Modernization Act, the secretary may approve a fast-tract product (intended for the treatment of a serious or life-threatening condition) upon determination that the product has an effect on a clinical endpoint or on a surrogate endpoint that is reasonably likely to predict clinical benefit (US Government Printing Office 1998). Approval may be subject to the requirement that the sponsor conduct appropriate postapproval studies to validate the surrogate endpoint.

From a regulatory perspective, a biomarker is not considered an acceptable surrogate endpoint for a determination of efficacy of a new drug unless it has been empirically shown to function as a valid indicator of clinical benefit. Many biomarkers will never undergo the rigorous statistical evaluation that would establish their value as a surrogate endpoint to determine efficacy or safety, but they can still have a use in drug development and regulatory decision making. Changes in biomarkers typically exhibit a time course that is different from changes in clinical endpoints and often are more directly related to the time course of plasma drug concentrations, possibly with a measurable delay. For this reason, exposure–response relationships based on biomarkers can help establish the dose range for clinical trials intended to establish efficacy. In some cases, these relationships can also indicate how soon titration should occur and can provide insight into potential adverse effects. Biomarkers can also be useful during the drug discovery and development stage, where they can help link preclinical and early clinical exposure–response relationships and better establish dose ranges for clinical testing (Guidance for Industry 2003). Examples for using biomarkers in drug discovery and preclinical development are summarized in Table 5.

The FDA Modernization Act states that confirmatory evidence, when combined with evidence from one adequate and well-controlled study, can support effectiveness as required for ordinary drug approvals (US Government Printing Office 1998). The quantity of evidence needed to support effectiveness, other than two adequate and well-controlled clinical trials, is discussed in Section II of the FDA Guidance for Industry (Food and Drug Administration. Center for Drug Evaluation and Research 1998). This guidance states that one adequate and well-controlled clinical efficacy study can sometimes be supported by evidence from a well-controlled study or studies using a pharmacologic effect, as a biomarker, that is not an established surrogate endpoint.

For regulatory acceptability of using biomarkers in discovery and preclinical and clinical drug development, discussion with the regulatory agency is beneficial to seek a common understanding and based on scientific rationale.

Table 5. Examples for using biomarkers in drug discovery and preclinical development

A. Biomarker in drug discovery		
	Aim	To identify a suitable target
	Example	Association of elevated serum cholesterol levels with an increased incidence of coronary heart disease provides an underlying rational search for developing new compounds that lower cholesterol
B. Biomarker in preclinical development (pharmacology)		
	Aim	To identify pharmacological activity
	Example	Association of blood pressure-lowering effect in suitable animal models with likelihood of having the intended therapeutic activity in patients
C. Biomarker in preclinical development (pharmacokinetics)		
	Aim	To guide dose selection and dose escalation strategies and to prevent adverse events
	Example	Association between blood levels and desired and undesired effects; PK/PD modeling
D. Biomarker in preclinical development (toxicology)		
	Aim	To identify potential risks
	Example	Drugs found to prolong the QT-interval in animals may warn of potential cardiovascular risk in subsequent clinical studies

3.6 Examples for Using Biomarkers in Early Clinical Development

A biomarker or surrogate marker is likely to be of greatest use if the therapeutic effect is difficult to measure, if there is a long delay between drug exposure and effect, if the novel drug acts on a pathway for which the role in disease is not well understood, or in conditions such as stroke, where large groups' sites may be required to demonstrate efficacy (Rolan et al. 1997, 2003).

Biomarkers are perhaps most useful in early phase of clinical development, when measurement of clinical endpoints may be too time-consuming or cumbersome to provide timely proof of concept (POC) or dose-ranging information. The use of changes in biochemical or clinical biomarkers in early clinical drug development to establish POC is only as good as the theoretical foundation for the biochemical or clinical biomarker. The scientific program for evaluating biomarkers must be planned as early as possible in the drug discovery and preclinical period of drug development with the aim to bring that biomarker into clinical trials and to establish the link between the biomarker and the clinical outcome.

3.6.1 Cardiovascular

Cholesterylester Transfer Protein Inhibition

A low level of high-density lipoprotein (HDL) cholesterol is the most common lipid abnormality observed in patients with known coronary heart disease and atherosclerosis; however, there are no therapies that substantially raise HDL cholesterol levels (Genest et al. 1991; Rader 2003). Inhibition of cholesterylester transfer protein (CETP) has been proposed as a strategy to raise HDL cholesterol levels (Tall 1993). In the guidelines set forth by the third Adult Treatment Panel of the National Cholesterol Education Program, a low HDL cholesterol level is defined categorically as a level below 40 mg per deciliter (Cleeman 2001). Gordon et al. (1989) have reported that an increase in HDL cholesterol by 1 mg/dl is associated with a 2%–4% reduction in the risk of cardiovascular events.

Using an ex vivo assay to measure the pharmacological activity of BAY 19-4789, a potent inhibitor of CETP, the inhibitory activity of the investigational drug could be demonstrated from the first clinical pharmacological studies in healthy volunteers (Schuehly et al. 2000). The first doses could already be discriminated by measurements of the CETP activity as a primary biochemical–pharmacodynamic parameter. The extent of inhibition correlated very well with the individual plasma concentrations. Sufficient decision-making information for the continued development of the compound was obtained from these early clinical–pharmacological studies in healthy volunteers.

Angiotensin II Receptor Antagonist

For drugs acting on the renin-angiotensin-aldosterone system (RAAS), vasoactive hormone levels (renin, angiotensin II, aldosterone) can be used as sensitive biomarkers in order to prove the mechanism of action (POM) in early phase I. In this case of an investigational angiotensin II receptor antagonist BAY 10-6734 from Bayer research (target indication: arterial hypertension), it was expected to see only minor effects on the arterial blood pressure of healthy subjects. Therefore, the effects of BAY 10-6734 on the vasoactive hormones were determined in parallel in order to allow POM and enable an early decision point for further development. In this crossover, double-blinded placebo-controlled study, BAY 10-6734 was administered to healthy volunteers in a dose range of 10–300 mg p.o. Plasma renin activity as the most immediate marker of the RAAS showed a dose-dependent, statistically significant increase even at the first investigated dose of 10 mg. Angiotensin II and aldosterone increases were moderate and only borderline statistically significant. Only the highest dosage group of 300 mg showed a slight decrease in blood pressure, which was accompanied by a compensatory increase in heart rate (Wensing et al. 2005). The pharmacodynamic properties of BAY 10-6734 in humans were quantitatively characterized from the rightward shifts to the agonist dose–response curves after administration of different single oral doses of the antagonist (Breithaupt-Grögler et al. 1997). The mechanism of action and the blood pressure-lowering effect could later be confirmed in clinical trials with patients suffering from mild to moderate hypertension (Wensing et al. 2005).

Sufficient decision-making information for the continued development of the compound was obtained from these early clinical–pharmacological studies in healthy volunteers and patients.

3.6.2 Pulmonology

Leukotriene Receptor Antagonist

Bronchial challenges are an ideal tool to assess the pharmacodynamic effect and duration of new antiasthmatic drugs early in the course of drug development and provide a rationale for the selection of doses and dosing schedules in clinical trials. Cysteinyl-leukotrienes appear to be of major importance in the pathophysiology of asthma. Inhalation of cysteinyl-leukotrienes leads to bronchoconstriction and can induce bronchial hyperreactivity both in healthy volunteers and the asthmatic (Smith et al. 1985; Adelroth et al. 1986; Arm et al. 1988). BAY x 7195 is a new and selective oral receptor antagonist of cysteinyl-leukotrienes which in preclinical studies has been shown to be effective against LTD_4 and antigen-induced bronchoconstriction in vitro and in vivo (Abram et al. 1987). Using a double-blind, placebo-controlled, crossover design, volunteers received 100, 250, and 500 mg of BAY x 7195 as tablet or 2 and 4 mg of BAY x 7195 aerosol, respectively Bronchoprovocation with nebulized LTD_4 was performed at different time points after drug administration. The specific airway conductance (SGaw) was used to assess the airway's response. Compared to placebo, the different doses of BAY x 7195 as tablets increased the concentration of LTD_4 needed to produce a 35% decrease in SGaw 2 h p.a. between 1- and 23-fold. Eight hours p.a., shifts in the concentration-response curve of between 1- and 13-fold could be observed (Wensing et al. 1994). Inhalation of 2 and 4 mg of BAY x 7195 significantly increased the concentration of LTD_4 needed to produce a 35% decrease in SGaw 20 min after drug administration between 4.5- and 149.2-fold. For both doses, only three out of six volunteers showed a protective effect against LTD_4-induced bronchoconstriction 8 h after drug administration (Wensing et al. 1996). For the different doses tested, a clear dose–response relationship was observed. Although the relevance of challenge studies with LTD_4 for the prediction of efficacy in bronchial asthma remains to be established, the outcome of bronchoprovocation testing in early studies in healthy volunteers is usually taken as a go/no go decision of a further drug development.

Phosphodiesterase 4 Inhibitor

Chronic pulmonary inflammation contributes to the underlying pathomechanism of asthma and chronic obstructive pulmonary disease (COPD). One concept could be to suppress inflammation in the lung by phosphodiesterase 4 (PDE_4) inhibition. Inflammatory activity of neutrophils is indicated by the formation and release of oxygen radicals upon stimulation with a bacterial peptide formyl-Met-Leu-Phe (f-MLP).

The measurement of oxygen radicals in neutrophils after ex vivo stimulation with f-MLP was used as a surrogate for the anti-inflammatory effects of BAY 19-8004, a selective PDE_4 inhibitor, in vitro and in vivo. When added to whole blood, BAY 19-8004 inhibited the oxygen radical formation with an IC_{50} value of 0.6 μM. Based on the findings that there was a significant inhibition at concentrations close to plasma levels achievable in clinical trials, this surrogate was also used in early phase

I studies of BAY 19-8004. The oxygen radical formation in neutrophils after ex vivo stimulation with f-MLP decreased in a dose-dependent manner with single oral dosages of BAY 19-8004 to healthy volunteers. After multiple oral dosing with 7.5 mg BAY 19-8004 to healthy volunteers, the inhibition was significant at the time of C_{trough} (Wandel et al. 2002). From these data, it appears reasonable to investigate the clinical benefit of BAY 19-8004 in such diseases whose pathomechanism is triggered by oxygen radical formation in neutrophils.

3.6.3 CNS System

For many CNS disorders, clinical trials are difficult and sometimes apparently adequately powered clinical trials may fail to show the known effect of a drug. When developing truly innovative therapies for such conditions, a model, even though imperfectly validated, can be used to support the decision to commit to a major clinical trial program and to assist with dose selection.

Pupillography is a noninvasive method which is used successfully in experimental investigations in animals as well as in investigations in healthy volunteers and patients (Böttcher 1999). Pupil reaction can be determined by light-evoked pupillography (LEP) to quantify pharmacodynamic effects on the autonomic and central nervous system. In order to investigate whether LEP can serve as a noninvasive biomarker for 5-HT_{1A} CNS effects, the influence of the 5-HT_{1A}-agonists ipsapirone and repinotan on initial pupil diameter (INIT) and reflex amplitude (RA) was investigated in 180 healthy volunteers and compared to body temperature and quantitative pharmaco-EEG. Both 5-HT_{1A} agonists showed a dose- and concentration effect-related reduction of the INIT and RA. LEP was more sensitive than body temperature or pharmaco-EEG (Böttcher et al. 2005). Therefore, pupillography should be a standard biomarker for effects of 5-HT_{1A} compounds.

These examples demonstrate the value of integrating biomarkers in the first exploratory studies, enabling an estimation of the risk of failure for a new drug already at an early decision point in clinical development.

3.7 Conclusions

A biomarker of drug effects should reflect a process on the critical path between the target identification, the pharmacological action of the drug and its effect on a disease. There is an acute need for effective biomarkers in every phase of research and development, from discovery to preclinical studies up to clinical trials in the early and late stages of development. Identifying novel biomarkers during these stages is an integral activity necessary for developing and delivering more efficacious, safer drugs to patient populations.

Whenever possible, appropriate biomarkers are incorporated into early clinical development studies to facilitate go/no go decision making, as well as providing information on efficacious dose range and exposure–response characteristics.

Success factors of biomarker development are:

- Team work involving individuals with diverse backgrounds
- Requisite core competencies
- Activities and processes aligned with existing discovery, preclinical, and early clinical development milestones
- Appropriate infrastructure, e.g., biomarker laboratories
- Mechanisms for prioritization of activities

When combined with measures of drug exposure and pharmacokinetics/pharmacodynamics (PK/PD) principles, biomarkers have a significant potential to facilitate critical decisions in the early phases of drug development. Expensive late-stage development failures could be shifted to less expensive early drug development go/no go decision making.

References

Abram TS, Cuthbert NJ, Francis HP et al (1987) Pharmacological profile of BAY x 7195, a structural antagonist of cysteinyl-leukotrienes. Am Rev Respir Dis 147 [Suppl]:A179

Adelroth E, Sterk P, Adelroth EC et al (1986) Airway responsiveness to leukotrienes C_4 and D_4 and to methacholine in patients with asthma and normal controls. N Engl J Med 315:480–484

Arm J, Spur W, Lee TH (1988) The effects of inhaled leukotriene E_4 in subjects with asthma and normal subjects. J Allergy Clin Immunol 82:654–660

Biomarkers Definition Working Group (2001) Biomarkers and surrogate endpoints: preferred definitions and conceptual framework. Clin Pharmacol Ther 69:89–95

Böttcher M (1999) Pupillography in clinical pharmacology. In: Kuhlmann J, Böttcher M (eds) Pupillography: principles, methods and applications. W. Zuckschwerdt Verlag, Clin Pharmacol 18:13–26

Böttcher M, Heinig R, Wensing G, Kuhlmann J (2005) Pupil reaction: a valid sensitive clinical biomarker for 5-HT1A compounds. Basic Pharmacol Toxicol 96:247

Breithaupt-Grögler K, Malerczyk C, Belz GG et al (1997) Pharmacodynamic and pharmacokinetic properties of an angiotensin II receptor antagonist – characterization by use of Schild regression technique in man. Int J Clin Pharmacol Ther 35:434–441

Cleeman JI (2001) Executive summary of the third report of the National Cholesterol Education Program (NCEP) Expert Panel on Detection, Evaluation, and Treatment of High Blood Cholesterol in Adults (2001) (Adult Treatment Panel III). JAMA 285:2486–2497

CMR International (2004) The CMR International R&D Factbook, Centre for Medicines Research International Ltd

Colburn WA (2000) Optimizing the use of biomarkers, surrogate endpoints, and clinical endpoints for more efficient drug development. J Clin Pharmacol 40:1419–1427

Danhof M, Alvan G, Dahl SG et al (2005) Mechanism-based pharmacokinetic – pharmacodynamic modelling – a new classification of biomarkers. Pharm Res 22:1432–1437

Food and Drug Administration Center for Drug Evaluation and Research, Center for Biologics Evaluation and Research (1998) Guidance for industry: providing clinical evidence of effectiveness for human drug and biological products. http://www.fda.gov/cber/guidelines.htm

Food and Drug Administration (2003) Guidance for industry. Exposure-response relationships – study design, data analysis, and regulatory applications. U.S. Department of Health and Human Services

Genest JJ, McNamara JR, Salem DN et al (1991) Prevalence of risk factors in men with premature coronary artery disease. Am J Cardiol 67:1185–1189

Gordon DJ, Probstfield JL, Garrison RJ et al (1989) High-density lipoprotein cholesterol and cardiovascular disease. Circulation 79:8–15

van den Haak MA et al (2004) Industry success rates (2004) – including a focus on decision outcomes. CMR04-234R

Kuhlmann J (1997) Drug research: from the idea to the product. Int J Clin Pharmacol Ther 35:541–552

Kuhlmann J (1999) Alternative strategies in drug development: clinical pharmacological aspects. Int J Clin Pharmacol Ther 37:575–583

Lesko LJ, Atkinson AJ (2001) Use of biomarkers and surrogate endpoint in drug development and regulatory decision making: criteria, validation, strategies. Annu Rev Pharmacol Toxicol 41:347–366

Lesko LJ, Rowland M, Peck CC et al (2000) Optimizing the science of drug development: opportunities for better candidate selection and accelerated evaluation in humans. J Clin Pharmacol 40:803–814

Rader DJ (2003) High-density lipoproteins as an emerging therapeutic target for atherosclerosis. JAMA 290:2322–2324

Rolan P (1997) The contribution of clinical pharmacology surrogates and models to drug development – a critical appraisal. Br J Clin Pharmacol 44:219–225

Rolan P, Atkinson AJ Jr, Lesk LJ et al (2003) Use of biomarkers from drug discovery through clinical practice: report of the Ninth European Federation of Pharmaceutical Sciences Conference on Optimizing Drug Development. Clin Pharmacol Ther 73:284–291

Schuehly U et al (2000) Randomized, double-blind, placebo-controlled, parallel group study on the safety, tolerability, pharmacodynamics and pharmacokinetics of BAY 19-4789 after oral administration of increasing single doses to healthy male subjects. (Internal report no. PH 30084, Bayer HealthCare AG)

Smith LJ, Greenberger PA, Patterson R et al (1985) The effect of inhaled leukotriene D_4 in humans. Am Rev Respir Dis 131:368–372

Tall AR (1993) Plasma cholesteryl ester transfer protein. J Lipid Res 34:1255–1274

US Government Printing Office (1997) Code of Federal Regulations, Title 21, Vol. 5, Part 314, Subpart H. US Government Printing Office, Washington, DC

US Government Printing Office (1997) Federal Food, Drug and Cosmetic Act. Sect. 505 (d)(5). US Government Printing Office,Washington, DC

US Government Printing Office (1998) Food and Drug Administration Modernization Act. Sect. 115. US Government Printing Office, Washington, DC

Wandel C et al (2002) The phosphodiesterase 4 (PDE_4) inhibitor BAY 19-8004 inhibits leukocyte oxygen radical formation (ORF) in vitro and in vivo. Clin Pharm Ther 71:P79

Wensing G, Heinig R, Priesnitz M, Kuhlmann J (1994) Effect of BAY x 7195, an oral receptor antagonist of cysteinyl-leukotrienes, on leukotriene D_4-induced bronchoconstriction in normal volunteers. Eur J Clin Pharmacol 47:227–230

Wensing G, Heinig R, Kuhlmann J (1996) Pharmacodynamics and pharmacokinetics of BAY x (7195) aerosol, a new and selective receptor antagonist of cysteinyl-leukotrienes, in normal volunteers. Br J Clin Pharmacol 42:171–178

Wensing G et al (2005) Early PoM/PoP in healthy volunteers or patients? Example of an angiotensin-II antagonist (abstract). 7th Congress of European Association for clinical Pharmacology and Therapeutics. Basic Clin Pharmacol Toxicol 97 [Suppl I]:26

Workshop on Bioanalytical Methods Validation for Macro-molecules: Summary Report 2001. Pharm Res 18:1373–1383

4 Using Exposure – Response and Biomarkers to Streamline Early Drug Development

J. Venitz

Abstract. Biomarkers (BMs) are biological measures of PD drug effects or disease markers that may represent clinically significant patient outcomes, either efficacy or toxicity. Their use in drug development, especially as an integral part of PK/PD modeling, has become a popular strategy for optimizing development time and resources. This approach supports quantitative integration of information across different species and throughout the clinical phases I–III. If the BM is based on the mechanism of action (MOA) of the drug, it is expected to follow an exposure–response relationship (E-R). If it is also involved in causal pathways in the pathophysiology of the disease (POD), it may become a surrogate marker (SM). SMs allow prediction of clinical outcomes for different dosing regimens of drug candidates and patient individualization of treatment in clinical practice. Appropriate evaluation of BMs by mechanistic, epidemiological, and clinical pharmacology studies as part of the drug development process allow scientists to establish clinically relevant ER. In early drug development, known ERs for BMs facilitate translation of in vitro findings to in vivo consequences, interspecies PK/PD comparisons, and streamlining of dose-finding phase I and II studies, as well as assessment of new dosing regimen candidates for their likely clinical efficacy and safety, extrapolation of clinical study results to special populations (e.g., pediatrics), and interpretation of exposure differences found in food, drug interaction and special populations studies. Recently, two novel BMs, namely, p_{50}, a measure of ex vivo/in vitro whole blood oxygen affinity and S_{pO2}, i.e., in vivo pulse oximetry, were used in the development of an allosteric synthetic hemoglobin modifier (SAM), efaproxiral, as PD endpoints; these BMs are based on the MOA of SAMs. Early use of these BMs established excellent in vitro/in vivo PK/PD correlations, appropriate interspecies PK and PD scaling as well as PD-guided phase I and II dose-finding studies. This approach allowed appropriate translation of in vitro and preclinical information along with early identification of sources of PK/PD variability. Frontloading drug development with the identification and use of mechanism-based (MOA/POD) BMs constitutes a rational strategy to quantitatively integrate PK/PD information and optimize dose finding.

Abbreviations

BM	Biomarker
CO	Clinical outcome
ER	Exposure–response relationship
Hb	Hemoglobin
MOA	Mechanism of action
PD	Pharmacodynamic(s)
PK	Pharmacokinetic(s)
POC	Proof of concept
POD	Pathophysiology of disease
PPB	Plasma protein binding

RBC	Red blood cell
RBCB	Red blood cell binding
SAM	Synthetic allosteric modifier
SM	Surrogate marker
TDM	Therapeutic drug monitoring

4.1 Guiding Principles

The effective and efficient use of biomarkers (BMs) and exposure–response relationships (ERs) has been proposed as a major strategy to streamline the drug development process by quantitatively integrating information about the clinical pharmacology of the investigational drug (Peck et al. 1992; Reigner et al. 1997; Derendorf at al. 2000; Lesko et al. 2000, Galluppi et al. 2000; Colburn 2000; Biomarkers Definition Working Group 2000; Down 2000; Lesko and Atkinson 2001; Venitz 2004).

4.1.1 Sources of Variability

Figure 1 illustrates the relationship between dosing regimen (route of administration, dose, dosing interval, etc.) and clinical outcomes (COs) and their respective consequences, namely benefit and harm (Venitz 2004). Pharmacokinetics (PK) relates the dosing regimen with systemic exposure, typically measured as plasma concentration of active moieties. Pharmacodynamics (PD) defines the relationship between drug exposure and pharmacological effect, measured as BM or surrogate marker (SM). Finally, therapeutics encompasses the pharmacological chain of events with the pathophysiology of disease (POD) to result in desirable and undesirable COs. Figure 2 shows how the POD leads from the disease initiation to measurable changes in SM to changes in COs, i.e., disease progression. These changes in SM provide diagnostic or prognostic information about disease progression and/or the therapeutic impact of the drug regimen (see Sect. 4.2.1). The overall clinical pharmacological paradigm in Fig. 1 emphasizes the importance of the optimal dosing regimen in achieving COs.

At each stage of this paradigm, i.e., PK and PD, variability within and between patients is introduced, which, for a given dosing regimen,

Fig. 1. The clinical pharmacology paradigm relating dosing regimens to clinical outcomes via drug exposure and response (biomarkers)

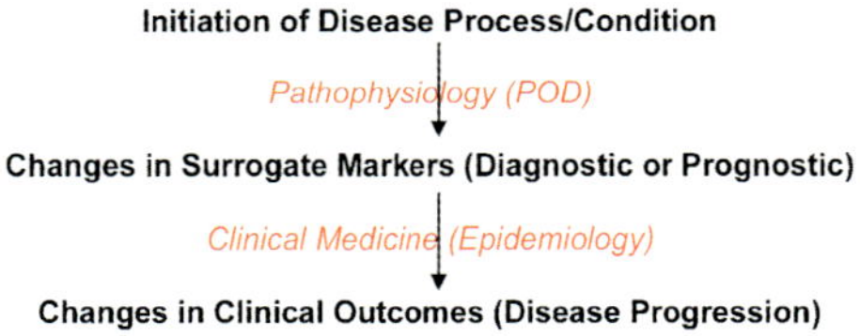

Fig. 2. Role of surrogate markers in the pathophysiology of disease

affects which, if any, CO is achieved, i.e., whether a patient is properly dosed, overdosed, or under-dosed. These sources of variability may be unexplained or explained. For example, intrinsic factors such as pharmacogenetics, gender, age, etc. or extrinsic factors such as concurrent medications or illnesses are clinical covariates (i.e., explained) that can impact significantly on the PK or PD of a drug, resulting in very different COs. On the other hand, formulation performance and medication adherence may contribute to the unexplained, residual variability in drug response. Therefore, early identification of potentially important clinical covariates as part of drug development will help in explaining the observed variability in drug response.

4.1.2 Exposure–Response Relationship

Figure 3 shows typical ERs, where with increasing drug exposure/dose, the fraction of the patient population showing efficacy or toxicity chang-

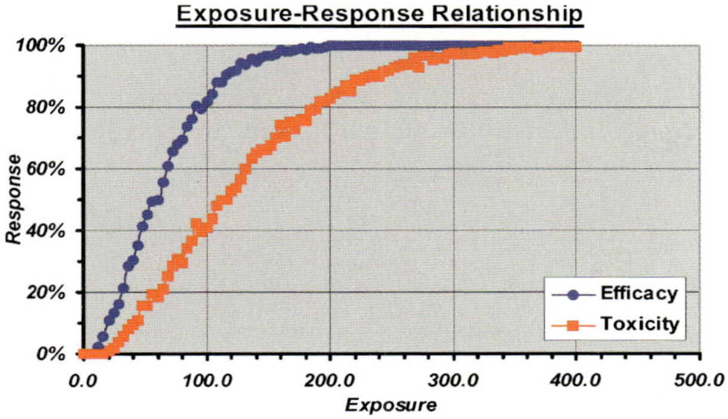

Fig. 3. Examples of exposure-response relationships for clinical efficacy and toxicity

es; typically, these relationships result in sigmoidal curves, i.e., after initial steep increases, they achieve a plateau. Obviously, in order for a drug to be therapeutically viable, the ER for toxicity has to be separated from the ER for efficacy by a right shift. The slope/sigmoidicity of these ERs is a direct consequence of the PK/PD variability discussed above. These relationships also illustrate that an optimal exposure (e.g., dosing regimen) is necessary to minimize the likelihood of toxicity while maximizing the likelihood of clinical efficacy. Therefore, knowledge of the underlying ER allows rational selection of an optimal exposure. Note that similar ERs are expected for BMs and SMs, but their shape may be quite different (Venitz 1994, 2004).

4.1.3 Clinical Dosing Regimens

As a result of the variability in PK/PD and the underlying ER for COs, selection of an optimal clinical dosing regimen may be essential for the success or failure of an investigational drug in the drug development process; the optimal dosing regimen for a given patient is designed to minimize the likelihood of toxicity and maximize the likelihood of efficacy. Clinical pharmacology information needs to be gathered and integrated to assess which of the following dosing regimens may be the most appropriate:

1. Fixed dosing regimen
   Every patient receives the same dose/dosing regimen, regardless of clinical covariate. This simple strategy may be an adequate dosing regimen for drugs with low toxicity potential, flat ER relationships, and/or low PK/PD variability between patients.
2. Individualized dosing regimens
   Patients may receive different dose/dosing regimens, depending on their individual needs. Two different strategies may be appropriate:

   2a. A priori dose adjustment
   The dosing regimen, typically the starting dose of a maintenance regimen, is based on clinical covariates that are known to be important sources of PK/PD variability. Fixed clinical covariates may include the patient's genotype, body weight, age, etc., while time-dependent covariates may include food, co-medications, or concurrent illnesses. The dosing algorithm relates the optimal dose to the value of the respective clinical covariate; the most common adjustment is dosing by body weight, i.e., mg/kg. This dosing strategy may be useful for drugs with known and measurable sources of variability, steep ERs, and possible significant benefits or harm.

   2b. A posteriori dose titration
   The patient is started on a dose (e.g., 2a), and markers are monitored that may predict the likelihood of COs; depending on these markers, the dosing regimen is dynamically adjusted over time. This kind of therapeutic drug monitoring (TDM) can be based on drug exposure (usually plasma concentrations), if PK is the major source of variability, or on BM or SM, if PD is the major source of variability. The dosing algorithm defines the marker to be monitored, its timing, and the actual dose adjustment. This is the most complicated clinical dosing regimen and is usually reserved for drugs with large PK/PD variability of unknown sources, steep ER relationships, and/or severe negative consequences of toxicity or lack of efficacy, i.e., a high risk of inadvertently over- or underdosing on a fixed dosing regimen. Example drugs would include cyclosporin and warfarin (Venitz 2004).

Prior to properly labeling and marketing a drug, the drug development process has to provide pertinent and comprehensive information to justify the optimal dosing algorithm used in the clinical or outpatient setting.

4.2 Role of Biomarkers in Drug Development

The drug development process includes the following major stages:

1. Preclinical studies include toxicology, pharmacology (mechanism of action, MOA), and PK/PD characterization in at least two non-human species.
2. Phase I includes exploratory dose ranging studies to assess safety, PK and PD, special population, drug–drug interaction and formulation development studies, typically in healthy subjects.
3. Phase II includes proof-of-concept (POC) studies to demonstrate the pharmacological and/or therapeutic concept of the investigational drug, as well as exploratory dose-ranging studies to assess preliminary efficacy and safety.
4. Phase III includes pivotal, large-scale clinical trials to confirm efficacy for the proposed indication, based on COs (see Sect. 4.2.1) along with clinical safety assessment.

In some cases, appropriate BMs can be measured throughout the preclinical and clinical development program and assist in identifying and separating PK and PD variability sources and help justify appropriate clinical dosing regimens.

4.2.1 Definitions

Consensus has been reached on the terminology of the different markers (Colburn 2000; Down 2000; Biomarkers Definition Working Group 2001):

1. Clinical outcome
   A clinically accepted indicator of disease state/progression, e.g., survival, morbidity, symptom scores. COs are associated with the clinical efficacy or safety/toxicity of a drug.

2. Surrogate marker
   A BM that predicts COs as accepted by the scientific, medical, and regulatory community. It may substitute for COs in the drug development process (dosing regimen and dosage form optimization and possibly drug approval) and in clinical medicine (TDM). At least some of the variability in COs is explained by changes in surrogate markers (Colburn 2000; Biomarkers Definition Working Group 2001).
3. Biomarker (intermediate endpoint)
   A biological (pathophysiological or pharmacological) indicator that can be measured as a result of a therapeutic intervention (drug). It may or may not be related to CO(s), but is involved in the chain of events in the POD and/or MOA the drug. Mechanism-based BMs are involved in the (presumed causal) chain of events in the MOA (PD marker; see Fig. 1) or POD (disease marker; see Fig. 2).

Note that drug toxicity may not be related to the MOA thought to be responsible for the therapeutic drug effect but rather be mediated by other (usually unknown mechanisms); therefore, safety BMs may be related to the mechanism of toxicity or be empiric and more difficult to anticipate.

Based on the (statistical) measurement scale on which they are measured, BMs can be classified as follows (Venitz 2004):

1. Graded response
   A quantifiable BM (such as an in vivo physiological response or in vitro test) that is causally and temporally linked to drug treatment and related to drug exposure (ER), e.g., blood pressure, serum cholesterol, International Normalized Ratio (for warfarin). These endpoints are usually chosen based on the MOA of the drug and known receptor-mediated physiological or biochemical responses.
   A graded response is a continuously scaled variable, can be measured repeatedly within the same individual, and is typically used for PK/PD modeling, particularly preclinically and in phases I and II.
2. Challenge response
   A quantifiable, graded response to a standardized exogenous challenge agent that is modified by administration of the drug of interest and related to drug exposure, e.g., exercise-induced tachycardia (to assess β_1-blocker activity), and histamine-induced bronchoconstric-

tion (to assess H_1-blocker activity). These markers are based on the MOA of the drug and sometimes on the POD. This kind of marker usually requires additional special clinical testing and is rarely used in clinical practice for dose adjustment.

A challenge response is a continuous variable (e.g., percent inhibition relative to baseline or placebo). It requires additional interventions, may not be repeated often within the same individual during a dosing interval, and contributes possibly unacceptable additional safety issues in early phase clinical drug development. However, it can be used for PK/PD modeling.

3. Categorical response

   A yes-or-no response due to drug administration that can be related to drug exposure, e.g., death, organ rejection, or incidence of adverse drug effects. This type of response is usually a clinically relevant outcome based on the disease progression in question, regardless of the MOA. It can be measured as part of clinical practice, but does not necessarily allow treatment adjustment.

   However, it can only be measured once within a given patient. It is a nominal variable that is not very informative statistically and requires a large sample size. It is used in phase II and III studies along with population PK/PD analysis.

4. Time-to-event response

   Time-to-event that is related to drug exposure, e.g., survival time or time to relapse. This type of response is usually a clinically relevant outcome based on the disease progression in question, regardless of the MOA. It can be measured as part of clinical practice, but usually does not allow treatment adjustment.

   It is a censored continuous variable that can only be measured once within a patient, is not very informative, and requires a large sample size in phase II and III studies along with population PK/PD analysis.

5. Event frequency/rate response

   Frequency of clinical events related to drug exposure, e.g., seizure frequency or frequency of cardiac arrhythmias.

   It is a censored continuous variable that can be measured more than once within a patient, but is not very informative and requires a large sample size in phase II and III studies and population PK/PD analysis.

4.2.2 Use of Biomarkers in Early Drug Development

Table 1 summarizes and compares the use of BMs throughout the drug development process:

In the preclinical development program, high drug exposures over long periods of time are achieved in two animal species (homogeneous, i.e., low PK/PD variability) as part of the toxicology package, which may allow monitoring of a BM for toxicity and safety along with its ER, while short-term dose-response and pharmacology studies based on the MOA may help in identifying and quantifying a putative BM for pharmacological POC and its ER.

Phase I studies in a very homogenous, healthy human population (i.e., low PK/PD variability) with high, short-term exposures and dose-ranging designs may assess the ER of BMs for safety and/or MOA to establish pharmacological POC.

Phase II studies in a somewhat homogenous target patient population (more PK/PD variability, clinical covariates) with high, short-term exposure and dose-ranging studies may assess BMs (MOA) or putative SMs (POD), their respective ER, as well as potentially important clinical covariates. Sometimes, for example, in the treatment of symptomatic diseases, COs may be evaluated in phase II as well, which allows the establishment of the ER for the efficacy CO.

Phase III studies in a large, heterogeneous target population with long-term drug exposures can help establish a correlation between BMs and COs to justify selection of SMs for efficacy and/or safety.

Table 1. Use of biomarkers across the drug development process

Phase	Population	Exposures	Endpoints
Preclinical	Two species	Long-term High exposures	E-R, BM (MOA) E-R, BM (safety)
Phase I	Healthy, homogenous	Short-term High exposures	E-R, BM (MOA) E-R, BM (safety)
Phase II	Target patients Homogenous	Short-term High exposures	E-R, BM/SM/CO (MOA/POD, safety)
Phase III	Target patients, less homogenous	Long-term	approvable SM, CO (POD, safety)
Phase IV	Target patients Representative	Long-term (High exposures)	SM/CO (POD, safety)

Finally, postmarketing studies in the heterogeneous, representative patient population at large with long-term and possible high exposures may allow correlation of possible SMs with COs based on POD, or empirically, for safety.

In early clinical drug development, BMs can be useful in translating information into the subsequent phase as follows:

1. Preclinical to phase I
   Useful preclinical information includes in vitro PK/PD (e.g., receptor binding and function studies, POD chain of intermediate events, drug metabolism, tissue binding, GI absorption) as well in vivo PK (absorption, distribution, elimination, differences between routes, dose, and species) and in vivo PD and ER (sensitivity, BM of safety/toxicity, putative BM of efficacy and biological activity). Quantitative methods (e.g., PK/PD modeling methods) can be used to translate and scale this preclinical PK/PD information to streamline the early phase I program (Peck et al. 1992; Venitz 1994, 1997; Reigner et al. 1997; Derendorf et al. 2000; Galluppi et al. 2001) by improving clinical study design:
   - Selection of the phase I, entry-into-humans, starting dose, dosing interval and dose escalation increment;
   - Selection of clinical endpoints to monitor such as drug exposures and BMs;
   - Selection of criteria for stopping dose escalation;
   - Identification of possible clinical PK/PD covariates (inclusion and exclusion criteria).
2. Phase I to phase II
   Early phase II studies include the POC study (pharmacology, MOA or disease impact, POD) as well as dose ranging studies to establish ER, while ancillary studies may assess food and drug formulation effects, special population studies, and drug–drug interaction studies. PK/PD information from phase I studies can be translated into useful phase II study design information as follows:
   - Selection of suitable starting dose, dose increments, and dosing intervals for phase IIa;
   - Rational selection of BM (putative SM) to monitor during dose escalation (phase IIa) and relate with COs (phase IIb);

- Identification of important clinical PK/PD covariates (inclusion and exclusion criteria in phase IIa, dose individualization in phase IIb);
- Evaluation of interindividual variability in drug exposure and ER for BM to separate PK from PD variability.

4.3 Case Example

Synthetic allosteric modifiers (SAMs) are synthetic compounds (clofibric acid derivatives) designed to bind to hemoglobin (Hb) and to reduce the allosteric equilibrium between deoxy- and oxy-Hb, leading to a release of oxygen (O_2) from Hb and increased tissue oxygenation. Efaproxiral (RSR13) is the lead compound of these SAMs and is currently undergoing phase III clinical testing as a radiation enhancer in oncology.

4.3.1 Ex Vivo and in Vivo Biomarkers of Synthetic Allosteric Modifiers

Two mechanism-based BMs (PD markers based on MOA) were identified during early preclinical SAM development:

p_{50} in whole blood
Ex vivo or in vitro measure of whole blood O_2 affinity that can be assayed by three-point tonometry or co-oximetry from whole blood samples. SAMs are expected to increase p_{50}, i.e., reduce whole blood O_2 affinity.

S_{pO2} In vivo measure of O_2 saturation of circulating blood, measured by pulse oximetry as a real-time BM. SAMs are expected to decrease S_{pO2}.

4.3.2 Utility of Biomarkers for Exposure–Response and Proof of Concept for Efaproxiral

During preclinical PK/PD and toxicological evaluation of efaproxiral, both BMs and drug concentrations in plasma and RBCs (red blood cells, MOA-target site) were evaluated for various IV dosing regimens to establish their respective ERs. The observed dose-limiting toxicity was related to an exaggerated PD effect (MOA), namely hypoxemia and hypoxia, predicted by changes in p_{50} and S_{pO2}. The in vivo and in vitro PK of efaproxiral was characterized by saturable plasma protein binding (PPB), saturable RBC binding (RBCB), as well as saturable renal tubular secretion and hepatic glucuronidation and biliary excretion (Venitz et al 1997; Hermening and Venitz 2000; Nasser et al. 2004). The in vivo ER was characterized by an instantaneous, linear PK/PD relationship between p_{50} and S_{pO2} and RBC concentrations of efaproxiral, parameterized by the slope, S_{p50}, a measure of intrinsic PD sensitivity (Venitz et al. 1997).

Interspecies comparison among rats, dogs, mini pigs, monkeys, and humans revealed that the systemic (plasma) clearance varied almost tenfold after correction for body weight, while volume of distribution varied fourfold after correction for body weight, resulting in sixfold differences in half-life; additionally, the RBC exposure relative to plasma exposure varied fivefold. Considering this high degree of PK variability, the interspecies PD variability of in vivo (and in vitro) S_{p50} was less than 20% (COV, $n = 4$; mean value, 0.02 mmHg/μg/ml), indicating that the major source of variability in BM response is due to PK, namely PPB, RBCB, and metabolism, while the intrinsic PD sensitivity is well conserved across species (Nasser et al. 2004).

Since both BMs were found preclinically to be predictive of dose-limiting toxicity and indicative of pharmacological POC (MOA), the phase I starting dose and dose escalation increment were based on p_{50} as BM. Furthermore, the phase I dose escalation study was BM-guided in its stopping rules, namely the achieved p_{50} was not to exceed 10 mmHg (BM-based dose escalation algorithm). Continuous safety monitoring by pulse oximetry (S_{pO2}) was done to assess for any exaggerated PD response as the main toxicity.

Figure 4 shows the time profile for plasma and RBC concentrations along with the changes in the two BMs following an infusion of 100 mg/kg of efaproxiral in a healthy volunteer (given as two 40-min infusions, separated by a 10-min washout). Both BMs show a transient change, peak at the end of the second infusion (i.e., at the time of peak plasma and RBC concentration), and return quickly to baseline (indicating a fully reversible drug effect). Figure 5 shows the corresponding ER for both BMs; as seen in the non-human species, a linear relationship

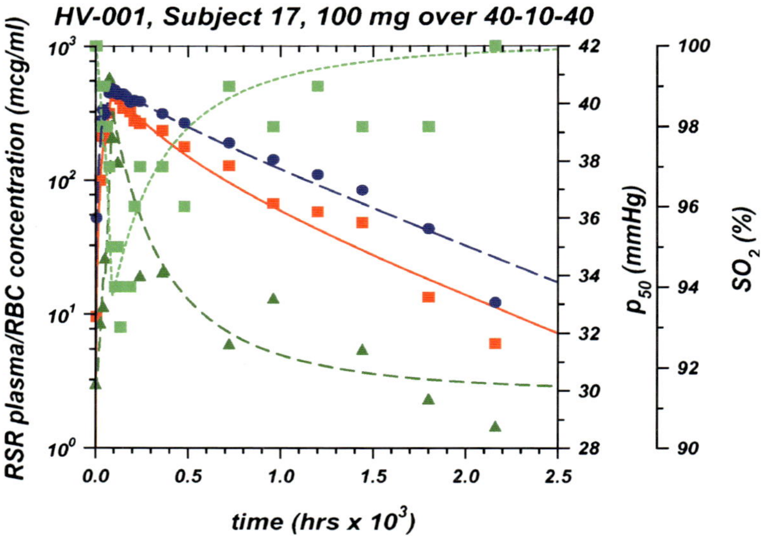

Fig. 4. Plasma and RBC concentration-time profile along with the BM (p_{50} and S_{pO2}) time profile following IV infusion of efaproxiral (100 mg/kg); *symbols* indicate observed values, *lines* indicate best fits following PK/PD modeling

with RBC concentration is apparent, with the slope, S_{p50}, as a measure of in vivo PD sensitivity. Using PK/PD modeling, appropriate PK/PD model parameters were estimated; as found preclinically, the intersubject PK variability in healthy human subjects ranged from 20% to 45% (COV, $n = 4$), while S_{p50} had a mean value of 0.02 mmHg/(μg/ml) and varied by less than 10%, confirming that the major source of variability is in PK, not intrinsic PD (Venitz et al. 1997).

These PK/PD findings were confirmed in phase II and phase III trials, and an optimal dosing algorithm was developed: given the high degree of PK variability (mainly unexplained), the presumed narrow therapeutic window, the severe potential consequences of toxicity (hypoxia) and, most importantly, the availability of two BMs predictive of toxicity, the proposed dosing algorithm involves initial, a priori dose adjustment based on body weight, followed by a posteriori titration for subsequent doses based on S_{pO2}, measured by widely available pulse oximetry.

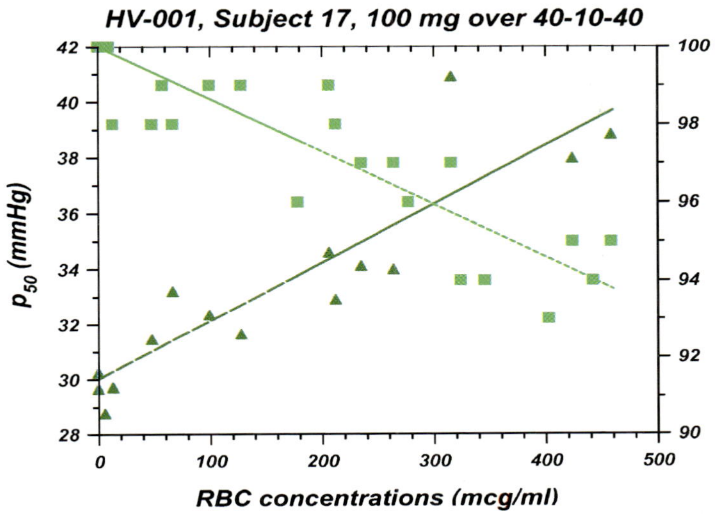

Fig. 5. Exposure–response relationship for both p_{50} and S_{pO2}; same patient as in Fig. 4

Overall, the early identification and systemic quantification of two mechanism-based BMs along with PK/PD modeling of their ER and their use throughout the preclinical and clinical drug development made it possible to quantitatively and incrementally integrate and model available in vitro and in vivo PK/PD information across species; streamline the phase I/II dose escalation scheme and safety monitoring procedures; provide early POC in humans (pharmacological POC-MOA, not therapeutic POC-POD); separate PK from PK/PD sources of variability; and design a rational, clinical dosing regimen using a BM as safety marker for ongoing phase III trials.

4.4 Conclusions

BMs have to be identified early during the drug discovery process and be evaluated systematically throughout the development process:

- During drug discovery, mechanistic understanding of POD and proposed MOA along with empiric evidence and theoretical models can identify candidate BMs.
- During preclinical development, in vitro receptor/enzyme binding or in vivo/ex vivo functional target testing can serve as BM; in vivo challenge paradigms should also be considered. Establishment of ER for BM allows rational interspecies scaling, selection of optimal dose and escalation increment for phase I studies, and identification of possibly important clinical PK/PD covariates.
- During early clinical development, phase I studies should include exploratory BM monitoring in this low-population-variability setting; in vivo challenge paradigms should be considered. Establishment of ER for BM allows rational phase II dose selection, identification of important clinical PK/PD covariates, possible correlation with COs (to possibly achieve SM status) and will assist in improved dose finding, i.e., help design and justify a dosing algorithm (i.e., fixed dose, a priori dose adjustment, a posteriori dose titration) for phase III studies and postmarketing practice.

Overall, using BMs throughout drug development will help to integrate information across development phases, identify sources of variability in assist in better dose selection and streamline drug development; however, the additional efforts in developing and implementing the BM assays have to be considered as well in assessing the utility of the BM approach.

References

Biomarkers Definitions Working Group (2001) Biomarkers and surrogate endpoints: preferred definitions and conceptual framework. Clin Pharmacol Ther 69:89–95

Colburn WA (2000) Optimizing the use of biomarkers, surrogate endpoints and clinical endpoints for more efficient drug development. J Clin Pharmacol 40:1419–1427

Derendorf H, Lesko L, Chaikin P, Colburn W, Lee P, Miller R, Powell R, Rhodes G, Stanski D, Venitz J (2000) Pharmacokinetic-pharmacodynamic modeling in drug research and development. J Clin Pharmacol 40:1–19

Down G (ed) Biomarkers and surrogate endpoints, 1st edn. Elsevier Sciences, Amsterdam, pp 1–9

Galluppi GR, Rogge MC, Roskos LK, Lesko LJ, Green MD, Feigal DW, Peck CC (2001) Integration of pharmacokinetic and pharmacodynamic studies in the discovery, development and review of protein therapeutic agents: a conference report. Clin Pharmacol Ther 69:387–399

Hermening A, Venitz J (2000) In-vitro glucuronidation of RSR13: correlation of in-vivo intrinsic hepatic clearance and metabolic interaction with probenecid. Clin Pharmacol Ther 67:133

Lesko LJ, Rowland M, Peck CC, Blaschke TF (2000) Optimizing the science of drug development: opportunities for better candidate selection and accelerated evaluation in humans. Pharm Res 17:1335–1344

Lesko LJ, Atkinson AJ Jr (2001) Use of biomarkers and surrogate endpoints in drug development and regulatory decision-making: criteria, validation, strategies. Annu Rev Pharmacol Toxicol 41:347–366

Nasser A, Joshi G, Venitz J (2004) Interspecies pharmacokinetic/pharmacodynamic differences and allometric PK scaling of RSR13, a novel synthetic allosteric modifier of hemoglobin. AAPS J 6:003062

Peck CC, Barr WH, Benet LZ, Collins J, Desjardins RE, Furst DE, Harter JG, Levy G, Ludden T, Rodman JH, Sanathanan L, Schentag JJ, Shah VP, Sheiner LB, Skelly JP, Stanski DR, Temple RJ, Viswanathan CT, Weissinger J, Yacobi A (1992) Opportunities for integration of pharmacokinetics, pharmacodynamics and toxicokinetics in rational drug development. Clin Pharmacol Ther 51:465–473

Reigner BG, Williams PEO, Patel IH, Steimer JL, Peck C, van Brummelen P (1997) An evaluation of the integration of pharmacokinetic and pharmacodynamic principles in drug development. Clin Pharmacokinet 33:142–152

Venitz J (1994) Pharmacokinetic-pharmacodynamic modeling of reversible drug effects. In: Derendorf H, Hochhaus G (eds) Handbook on pharmacokinetic-pharmacodynamic correlations, 1st edn. CRC Press, Boca Raton, FL pp 1–34

Venitz J (1997) Prerequisites, objectives and clinical trial designs of a phase I program. Pharmacokinetic/pharmacodynamic analysis II: accelerating drug discovery and analysis. International Business Communications, Southborough, MA

Venitz J, Slattum PW, Gerber M, Abraham D (1997) Pharmacokinetic-pharmacodynamic modeling of the effects of an allosteric hemoglobin modifier, RSR13, in healthy young volunteers. Clin Pharmacol Ther 61:154

Venitz J (2004) Surrogate markers in drug development. In: Sahajwalla CG (ed) New drug development: regulatory and scientific principles for clinical pharmacology and biopharmaceutics. CRC Press, Boca Raton, FL, pp 213–228

5 Experiences with Dose Finding in Patients in Early Drug Development: The Use of Biomarkers in Early Decision Making

S.R. Sultana, S. Marshall, J. Davis, B.H. Littman

Abstract. With the increasing cost and complexity of drug development, biomarkers will play an increasing role in the early phases. Biomarkers can be classified into target, mechanistic, or outcome with varying degrees of linkage to disease or treatment effect. They can be used to determine proof of concept by characterising the efficacy or safety profiles, or determining differentiation from any competitor drugs. PK/PD modelling of biomarker data for novel and marketed compounds can be used to predict outpatient dose response. Subsequent simulations may replace or reduce the size and cost of larger phase 2b outpatient studies. Two examples of biomarkers and PK/PD modelling used to characterise dose response are presented. Penile plethysmography (RigiScan Plus) in male erectile dysfunction and phenylephrine challenge urethral pressure in benign prostatic hyperplasia are used to reduce time and cost to reach major exploratory development decision points in these indications.

5.1 Introduction

In the conventional drug development paradigm, a new chemical entity (NCE) undergoes evaluation of pharmacokinetic, safety, and toleration profiles in phase 1 studies in healthy volunteers, followed by efficacy assessment in phase 2a outpatient studies using registration endpoints, before progressing to larger dose response and confirmatory outpatient studies, again with conventional endpoints. The pharmaceutical industry is faced with increasing costs of drug development, higher regulatory hurdles prior to drug approval, and erosion of the period of exclusivity by generic drug manufacturers challenging a variety of patents (Grabowski 2004). These factors combine to force many pharmaceutical companies to rethink their drug development strategies and increasingly depend on biomarkers of efficacy and novel study designs to reach major investment decision points quickly and cost effectively.

Most research and development organisations have drug portfolios that include a mix of precedented ('me too') mechanisms as well as novel, unprecedented targets. There is a gradual change in the R&D landscape with a greater proportion of novel targets. The emphasis on unprecedented mechanisms is expected to rise as drug developers capitalise on the better understanding of disease and metabolic pathways as more genomic, proteomic and metabonomic data become available. With the lower success rate of unprecedented mechanisms, the main focus with these is to quickly establish whether there is meaningful efficacy at safe and well-tolerated doses. Ideally this should be achieved in early development through the use of biomarkers to determine proof of concept (POC).

5.2 Use of Biomarkers in Early Decision Making

The rising cost of developing new drugs and shortening period of exclusivity requires smarter clinical programs that allow quick, cost-effective and well-informed decision making. This can be achieved through the sensible use of biomarkers to achieve POC and to explore the effective dose range prior to large-scale outpatient studies.

A biomarker is a biological measure that can detect physiological changes due to a disease process or therapeutic intervention. Biomarkers can be used as diagnostics to identify target populations and characterise disease severity, or as measures of drug efficacy, safety, or differentiation. A surrogate endpoint is a biomarker accepted by regulatory agencies as a substitute for a clinical endpoint (e.g. HIV load for the stages of HIV/AIDS, LDL lowering for the risk of coronary artery disease, blood pressure lowering for the incidence of stroke, and haemoglobin $A1_c$ for the control of diabetes). It is important to note that for the purpose of decision making in early drug development, there is no need to develop and validate each biomarker to the level of regulatory acceptance as a surrogate endpoint. The validation process should be fit for purpose to give sufficient confidence that the biomarker performs reliably in reducing uncertainty prior to large clinical trials.

Biomarkers can be broadly classified into target, mechanism or outcome categories. A target biomarker is one that measures a direct pharmacological effect as a result of an interaction with the target receptor, enzyme or transport protein (e.g. elevation of substrate levels with enzyme inhibition). A mechanism biomarker is one where the downstream pharmacological effect measured is directly related to the expected mode of action of the drug (e.g. measurement of changes in vaginal blood flow or lubrication for a vasoactive approach used in the treatment of female sexual arousal disorder). An outcome biomarker is one that substitutes for a clinical efficacy or safety outcome and is clearly linked with clinical benefit regardless of the mechanism of action of the drug (e.g. blood pressure changes in hypertension). A target or mechanism biomarker can also be linked to outcome but is specific for that mechanism of action. Hence, it may not be appropriate for other mechanisms that treat the same condition. Biomarkers are also described with regards to the degree of linkage with the disease process, and efficacy or safety in the wider population. Hence high linkage denotes a biomarker where changes are shown to correlate reproducibly with disease state or treatment effect. Conversely, low linkage may reflect either data showing a poor correlation or lack of information on clinical outcome (i.e. an immature technology).

Besides their role in early decision making, biomarkers are also useful for characterising a drug's mode of action. This is often important for novel mechanisms or new indications to help explain a drug's beneficial effects to regulatory agencies. Biomarkers can also be used to define additional characteristics of the NCE such as time to onset and offset of activity (and hence help define the dosing regimen required), features that are often poorly characterised with conventional endpoints in outpatient studies. Biomarkers may also be useful in defining efficacy in special populations, determining differentiation from competitor drugs, and identifying patients that respond to treatment to allow enrichment of subsequent larger studies.

5.3 Characterising the Dose Response

The dose response in any given indication is usually determined in a large phase 2b outpatient study using conventional endpoints. The ultimate goal would be to use a biomarker-based study design to replace the larger study. This may be achieved by using an outcome biomarker that has high linkage to outpatient effect. A mechanism or target biomarker with lower linkage to outcome can still be used in early studies to describe the pharmacological concentration response relationship and narrow the range of doses that need to be assessed in the phase 2b study, reducing the size of this study and hence saving time and money.

Pharmacokinetic/pharmacodynamic (PK/PD) modelling can used to quantify the extent of linkage between preclinical animal model and clinical biomarker data, and conventional outpatient endpoints. Modelling for emerging NCE data is based on reference animal model and biomarker data generated on marketed drugs whose outpatient dose response is well understood.

PK/PD modelling is used to characterise the time course relationship between drug dose, plasma concentration, pharmacokinetic profile, drug effect or side effects and relevant covariates at each stage of drug development. The use of PK/PD modelling increases the accuracy of interpretation of both preclinical and clinical data and hence allows greater precision in predicting the dose response. It therefore enhances the use of biomarker data to replace or reduce the size of phase 2b studies.

5.4 Applied Clinical Biomarkers: Two Examples from the Genitourinary Therapeutic Area

The penile plethysmography (RigiScan Plus) technique in male erectile dysfunction and the phenylephrine challenge urethral pressure methodology in benign prostatic hyperplasia will be discussed to illustrate how biomarkers can be used to determine POC and to characterise dose response early in a drug development program.

5.4.1 Penile Plethysmography Technique

Male erectile dysfunction (MED) is defined as the inability to achieve or maintain a penile erection sufficient to permit satisfactory sexual intercourse and can be due to psychogenic, vascular, and/or neurogenic disorders (NIH Consensus Conference 1993). It is estimated that MED affects 10–30 million men in the US and more than 140 million men worldwide (Furlow 1985; Kaiser 1999).

The RigiScan Plus technique was developed as a diagnostic technique to help confirm erectile dysfunction as well as to distinguish between psychogenic and organic causes. It has since been used to characterise the efficacy of drugs for MED and has been shown to be useful for both centrally acting drugs and peripheral vasodilatory mechanisms (Boolell et al. 1996; Diamond et al. 2004; Heaton et al. 1995; Pryor 2002). In this technique, sensor loops are placed around the tip and the base of the penis. The loops are connected to a recording device that monitors the circumference (tumescence) and radial tension (rigidity) at tip and base of the penis (Fig. 1). Patients who experience 60% base rigidity in this lab-based setting will achieve an erection sufficient for sexual intercourse in an at-home environment (Kaneko and Bradley 1986; Ogrinc and Linet 1995).

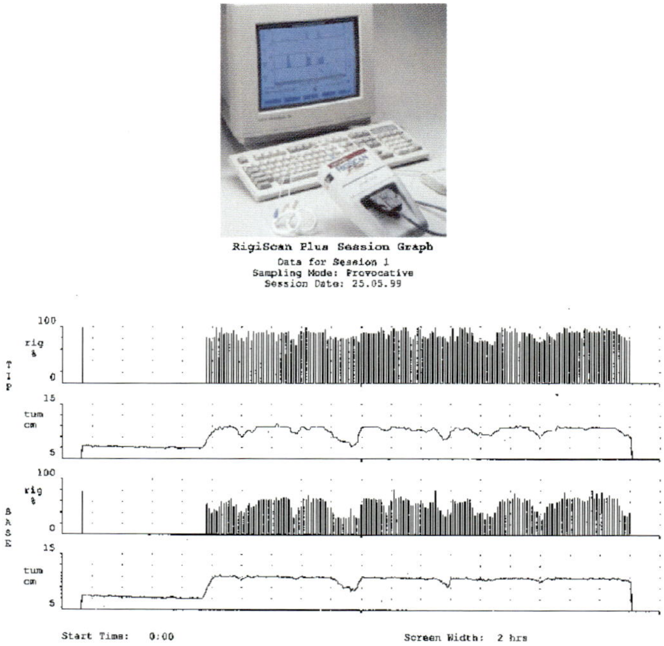

Fig. 1. RigiScan Plus equipment and representative trace from VSS session

Penile plethysmography and outpatient data for sildenafil (10–100 mg) demonstrate a good association between erectile activity as determined by the lab-based assessment and eventual efficacy in the outpatient studies (Figs. 2, 3). PK/PD modelling has shown a high degree of linkage between the lab-based efficacy assessment and outpatient conventional endpoints. This is borne out by similar reports of efficacy in penile plethysmography studies, subsequently translating into outpatient effect with other phosphodiesterase Type V inhibitors (Pryor 2002; Porst 2002) as well as with a centrally acting mechanism (Heaton 2000). The RigiScan Plus technique has therefore been validated as an outcome biomarker and can be used to define efficacy and dose response in MED patients. It is used to define POC in this indication by defining the therapeutic index for new compounds. This is exemplified below.

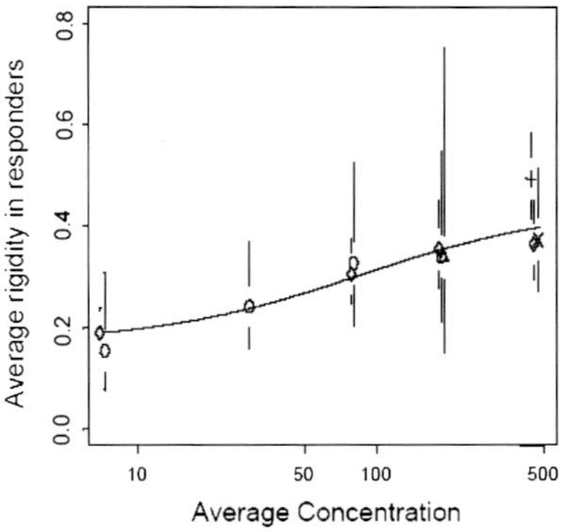

Fig. 2. Sildenafil penile plethysmography concentration response. Average rigidity (proportional scale; 1 = 100% rigidity) of penile erection during VSS period

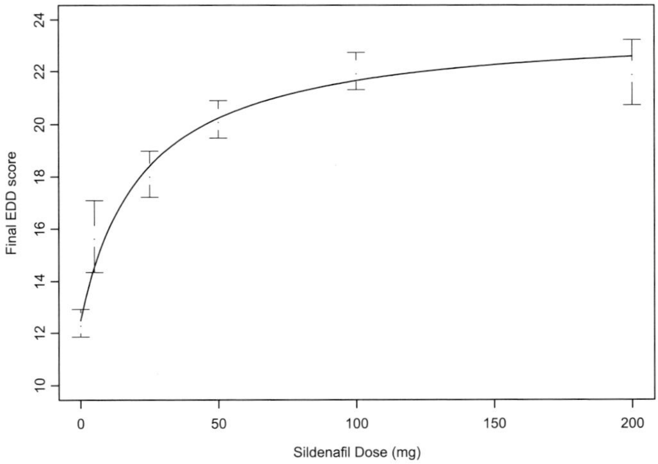

Fig. 3. Sildenafil outpatient study dose response. Conventional endpoint used is questionnaire-based erectile dysfunction domain (EDD) score of International Index of Erectile Function

CMPD-A is an agent effective in preclinical models of erectile function intended for the p.r.n. treatment of MED. This compound was assessed in a randomised, double blind, placebo-controlled and single dose four-way crossover study in MED patients. Each subject received two dose levels of CMPD-A, placebo and sildenafil 100 mg in a randomised fashion. The study assessed the efficacy of a 40-fold dose range of CMPD-A in an incomplete block design in 24 subjects. The primary endpoint was duration of erections of 60% rigidity at the base of the penis, recorded in MED patients during 90 min of visual sexual stimulation (sequences of erotic videos). Blood samples for drug concentration were taken at the end of the penile plethysmography period.

This technique showed that all doses of CMPD-A tested were efficacious relative to placebo (Fig. 4). PK/PD modelling of exposure and penile plethysmography data showed a good concentration–response relationship (Fig. 5). The modelling of this data against prior data for sildenafil allowed estimation of the efficacious dose range relative to sildenafil. This was used to scale the outcome dose response of sildenafil to predict what it would be for CMPD-A.

Simulations based on the predicted outcome dose response that was informed by the penile plethysmography study were used to inform the choice of doses and number of subjects of a subsequent outpatient study. This biomarker approach coupled with PK/PD analysis helped to reduce the size, and hence cost and duration, of the Phase 2B study. The RigiScan Plus technique was also used to determine time to onset

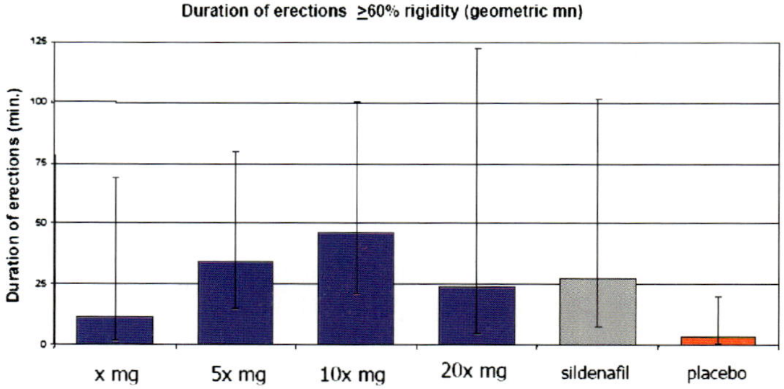

Fig. 4. Penile plethysmography study in MED patients. Efficacy of CMPD-A compared to sildenafil and placebo. Duration of erections of $> 60\%$ rigidity during period of VSS

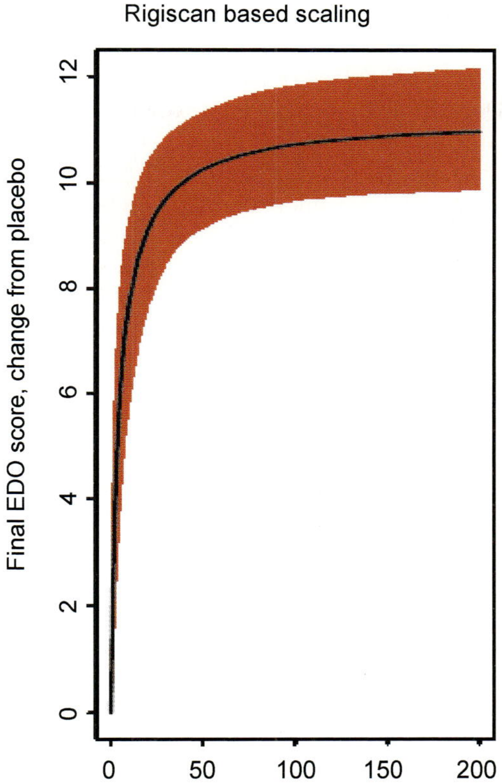

Fig. 5. PK/PD modelling of penile plethysmography data for CMPD-A. Scaling relative to predicted outpatient conventional endpoint (EDD score) improvement. Mean and 95% CI of predictions presented

of activity and there are plans to use this methodology to characterise time to offset in a separate study. These data will help inform dosing instructions and design of Phase 3 studies.

5.4.2 Phenylephrine Challenge Urethral Pressure Technique

Benign prostatic hypertrophy is an age-related condition that affects a significant proportion of men. Its symptoms reflect an increasing difficulty with micturition as well as bladder overactivity as a result of adaptive changes in the detrusor muscle. Treatments are categorised into (a) those that relax the smooth muscle component of the prostatic stroma, leading to a lowering of urethral resistance and hence improvement in flow (e.g. alpha-adrenoceptor antagonists), and (b) those that cause a reduction in prostate size by inhibiting growth (e.g. 5α-reductase inhibitors). Alpha-blockers have a number of dose-limiting side effects, particularly postural hypotension. Selective $\alpha 1_A$-adrenoceptor antagonists have an improved therapeutic index but are still limited by blood pressure effects.

The urethral pressure challenge methodology was designed to assess drugs that target urethral prostate tone, i.e. cause smooth muscle relaxation and lowering of urethral resistance. The technique makes use of custom-designed 9-Fr urethral catheters (Gaeltec Ltd, Dunvegan, UK) that contain microtip pressure transducers. After insertion into the urethra, the catheters are anchored at the bladder neck via an incorporated inflatable balloon (Fig. 6). These catheters have been used to record urethral pressure continuously for up to 6 h at a time in healthy male volunteers (aged 40–65 years). Simultaneous blood pressure recording allows a direct comparison of the effects of drugs on urethral vs blood pressure. Early studies showed that changes in baseline urethral pressure over a 4- to 6-h period were less reliable than intermittent escalating infusions of phenylephrine, an alpha-adrenoceptor agonist that elevates urethral and blood pressures. By using the phenylephrine challenge, these early studies showed that the methodology was able to determine an alpha-blocker effect on urethral and blood pressures, estimating selectivity for the urethral effect, and therefore the therapeutic index relative to hypotensive changes (Sultana et al. 1998). This methodology has since been used to determine the urethral vs blood pressure effect of novel agents that cause prostatic smooth muscle relaxation, as exemplified below.

CMPD-B is an agent that is effective in in vitro tests and an in vivo animal model of prostatic smooth muscle relaxation with selectiv-

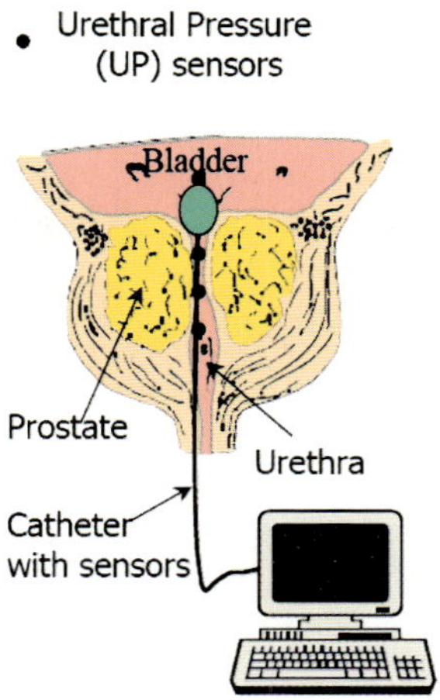

Fig. 6. Diagrammatic representation of urethral pressure catheter in place. Coronal section through male bladder base and prostate with self-retaining catheter in situ. The catheter has one bladder pressure sensor and three prostatic urethra sensors. Urethral pressure reading is taken from middle prostatic urethral sensor

ity for prostatic vs blood pressure changes in this animal model. The effect of this compound (tenfold dose range) on urethral and blood pressures was assessed in two studies involving a total of 22 healthy male volunteers aged 40–65 years. The studies were randomised, double blind, placebo-controlled, four- and five-way crossover designs and included tamsulosin 0.4 and 0.8 mg as reference doses. Phenylephrine dose-escalating infusions were administered at the estimated times of peak plasma concentrations for CMPD-B and for tamsulosin. Blood samples were taken during the study days to determine the pharmacokinetic profile of CMPD-B.

PK/PD modelling of the phenylephrine effect on urethral and blood pressures allowed estimation of the effect of CMPD-B relative to tamsulosin. The relative therapeutic index of CMPD-B compared to tamsulosin was estimated. Figure 7 shows that CMPD-B has greater urethral selectivity relative to hypotensive effect compared to tamsulosin. The relative urethral effect measured was used to scale outcome data available for tamsulosin to make predictions for CMPD-B. This was subsequently

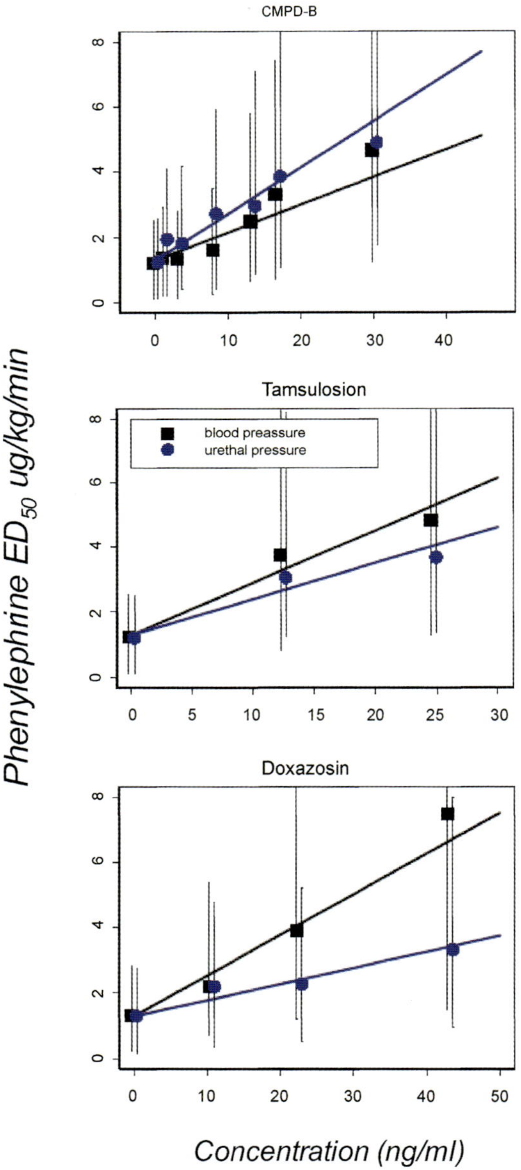
CMPD-B
Tamsulosion
blood preassure
urethal pressure
Doxazosin
Phenylephrine ED50 ug/kg/min
Concentration (ng/ml)
0
2
4
6
8
0
10
20
30
40
0
5
10
15
20
25
30
0
10
20
30
40
50

Fig. 7. Relative effect of CMPD-B and of tamsulosin on urethral and blood pressures. Phenylephrine ED_{50} is used to calculate pressor effect at each dose level of the agents. This is calculated as the dose of phenylephrine that causes a 50% elevation of urethral pressure and blood pressure relative to the maximal placebo rise observed

confirmed in a large outpatient Phase 2B study in benign prostatic hypertrophy patients (Fig. 8). This technique therefore allows a comparison of the effect of novel agents relative to the current market leader tamsulosin in a POC study conducted in healthy volunteers. The technique can be used to see whether a novel agent is able to fulfill the product profile of superior efficacy with minimal blood pressure effects that is required in this indication.

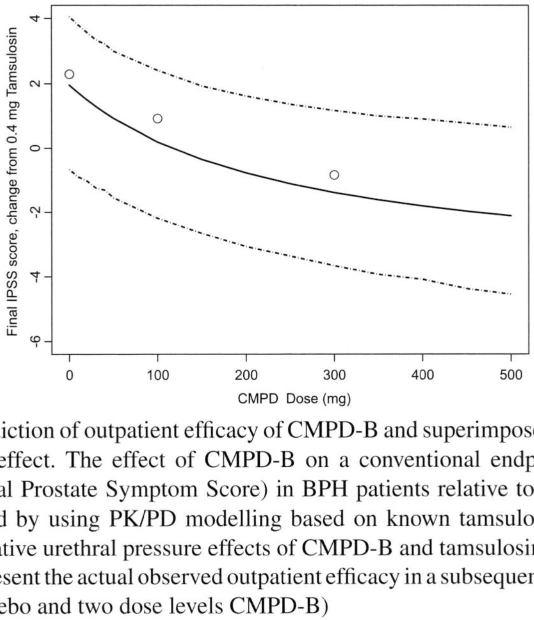

Fig. 8. Prediction of outpatient efficacy of CMPD-B and superimposed observed outpatient effect. The effect of CMPD-B on a conventional endpoint (IPSS, International Prostate Symptom Score) in BPH patients relative to tamsulosin is estimated by using PK/PD modelling based on known tamsulosin efficacy and the relative urethral pressure effects of CMPD-B and tamsulosin. The *three points* represent the actual observed outpatient efficacy in a subsequent outpatient study (placebo and two dose levels CMPD-B)

5.5 Summary

We are faced with more expensive clinical development programs and an increasingly risk-averse regulatory environment. Consistent use of large outpatient studies to define POC is not likely to be sustainable, particularly with unprecedented targets. There is therefore an increased pressure to use smart clinical trial designs in exploratory development. Biomarkers should be developed and used where appropriate to replace larger, conventional endpoint studies. At the very least, these biomarkers should better inform the design of outpatient studies to reduce the risk of failure and allow smaller studies that assess fewer doses.

Clinical biomarker studies are becoming an essential aspect of early drug development, with key decisions based on their outcome. In this new paradigm, PK/PD modelling has an important role in using prior data on reference drugs to quantify the degree of linkage between preclinical animal models, clinical biomarkers and outpatient response. Predictions of the likely outpatient dose response can then be made on the basis of the emerging biomarker data for novel compounds. Together, these approaches provide a powerful tool in guiding dose selection and optimal outpatient trial design for novel compounds.

The examples cited in this paper should convince the reader of the important role biomarker methodologies could play in early drug development. We now routinely use the RigiScan Plus technique in male erectile dysfunction and have found the phenylephrine challenge urethral pressure technique equally useful in benign prostatic hypertrophy. Besides their use in determining POC, these techniques can also be used to characterise the dose response, as well as additional pharmacodynamic properties such as time to onset and offset of activity. By allowing direct comparison with competitor agents, these techniques enable the differentiation of novel compounds to be determined. Techniques such as RigiScan Plus can also be used to define efficacy in special populations and therefore help in the design of phase 3 programs. Use of these biomarker techniques leads to more efficient drug development programs, with early go/no go decisions reached in a cost-effective manner.

Acknowledgements. We would like to acknowledge Dr. Jaap Mandema (Pharsight Corp now Quantitative Solutions) for the modelling and simulation work he did for the examples presented.

References

Boolell M, Gepi-Attee S, Gingell JC et al (1996) Sildenafil, a novel effective oral therapy for male erectile dysfunction. Br J Urol 78:257–261

Diamond LE, Earle DC, Rosen RC et al (2004) Double-blind, placebo-controlled evaluation of the safety, pharmacokinetic properties and pharmacodynamic effects of intranasal PT-141, a melanocortin receptor agonist, in healthy males and patients with mild-to-moderate erectile dysfunction. Int J Impot Res 16:51–59

Furlow WL (1985) Prevalence of impotence in the United States. Med Aspects Hum Sex 19:13–16

Grabowski H (2004) Are the economics of pharmaceutical research and development changing? Productivity, Patents and political pressures. Pharmacoeconomics 22 [Suppl 2]:15–24

Heaton JPW, Morales A, Adams MA et al (1995) Recovery of erectile function by the oral administration of apomorphine. Urology 45:200–206

Heaton JPW (2000) Apomorphine: an update of clinical trial results. Int J Impot Res 12 [Suppl 4]:S67–S73

Kaiser FE (1999) Erectile dysfunction in the aging man. Med Clin N Am 83:1267–1278

Kaneko S, Bradley WE (1986) Evaluation of erectile dysfunction with continuous monitoring of penile rigidity. J Urol 136:1026–1029

NIH Consensus Conference (1993) Impotence. NIH Consensus Development Panel on Impotence. JAMA 270:83–90

Ogrinc FG, Linet OI (1995) Evaluation of real-time RigiScan monitoring in pharmacological erection. J Urol 154:1356–1359

Porst H (2002) IC351 (tadalafil, Cialis): update on clinical experience. Int J Impot Res 14 [Suppl 1]:S57–S64

Pryor J (2002) Vardenafil: update on clinical experience. Int J Impot Res 14 [Suppl 1]:S65–S69

Sultana SR, Murray K, Craggs MD et al (1998) Alpha, adrenoceptor antagonist selectivity determined by simultaneous blood pressure and long-term urethral pressure monitoring in men. Br J Clin Pharmacol 45:192

6 Genotype and Phenotype Relationship in Drug Metabolism

I. Roots, G. Laschinski, F. Arjomand-Nahad, J. Kirchheiner,
D. Schwarz, J. Brockmöller, I. Cascorbi, T. Gerloff

Abstract. Pharmacogenetics, one of the fields of clinical pharmacology, studies how genetic factors influence drug response. If hereditary traits are taken into account appropriately before starting drug treatment, the type of drug and its dosage can be tailored to the individual patient's needs. Today, the relationships between dosage requirements and genetic variations in drug-metabolizing enzymes such as cytochrome P450 (CYP) 2D6, CYP2C9, and CYP2C19 or in drug transporters such as p-glycoprotein (*ABCB1*) and OATP-C (*SLC21A6*) are substantiated best. A standard dose will bring about more adverse effects than usual if enzymatic activity is lacking or feeble. Sometimes, however, therapeutic response might be better because of higher concentrations: proton pump inhibitors for eradication of *Helicobacter pylori* are more efficacious in carriers of a deficient CYP2C19 variant. In some cases, genetic tests can help distinguish between responders and nonresponders of a specific drug treatment, and genotype-based dosage is possible.

6.1 Introduction

Pharmacogenetics, a subdiscipline of clinical pharmacology, seeks to optimize drug treatment by tailoring drug selection and drug dosage to the patient's genetic make-up. This concept does not present the physician with new chemical entities but helps streamline the therapeutic regimen. Drug treatment is more efficacious if the patient gets individually optimized dosages of only those medicines to which he responds. It is also safer because some side-effects are avoided (Fig. 1).

Pharmacogenetics has greatly profited from the progress in molecular biology in the 1990s and, above all, from the Human Genome Project. Nevertheless, some pharmacogenetic phenomena had already been described years ago, e.g., hemolytic anemia occurring in carriers of glucose-6-phosphate-dehydrogenase deficiency who had taken specific drugs or food, polyneuropathy in slow acetylators following isoniazid treatment (Bönicke and Lisboa 1957), and prolonged apnea when succinylcholine was administered to carriers of a specific cholinesterase variant (Kalow 1956).

There are many examples of genetic variations in drug receptors and drug targets with important clinical consequences. However, drug-

- Drug dosage
- Drug selection
- Responders / Non-responders
- Side effects
- Disease susceptibility – Prevention
- Pharmacovigilance

Fig. 1. Topics of individualized drug treatment and pharmacogenetics metabolizing enzymes are the most likely reason for interindividual variation in drug response. Genetic variations have been identified in nearly every important enzyme that is involved in the metabolism of xenobiotics (Tables 1, 2). Their functional consequences range from a moderate reduction to a complete lack of enzymatic activity, but even a consider-

Table 1. Hereditary polymorphisms in important enzymes metabolizing drugs and xenobiotics and their therapeutic consequences (Roots et al. 2004)

Phase-I enzymes	Frequencies of genetic variants[a]	Examples of affected drugs
CYP1A2	Europeans: 46% highly inducible	Acetaminophen, caffeine, clozapine, imipramine, lidocaine, theophylline
CYP2A6	Europeans: 1% activity reduced	Fadrazole, losigamone, halothane, nicotine, tegafur
CYP2B6	Europeans: approx. 2% activity reduced	Bupropion, propofol
CYP2C8	Europeans: approx. 1.7% activity reduced	Carbamazepine, cerivastatin, paclitaxel, pioglitazone, rosiglitazone, verapamil
CYP2C9	Europeans: 1%–3% activity reduced	Celecoxib, clopidogrel, diclofenac, fluvastatin, glibenclamide, ibuprofen, losartan, nateglinide, phenprocoumon, phenytoin, piroxicam, sildenafil, tolbutamide, torasemide, warfarin
CYP2C19	Europeans: 3% activity lacking Asians: 14%–20% activity lacking	Cyclophosphamide, diazepam, lansoprazole, omeprazole, pantoprazole, proguanil, propranolol, rabeprazole
CYP2D6	Europeans: 7% activity lacking Far East Asians: 1% activity lacking. Europeans: 2%–3% activity very high Arabs and Ethiopians: 5%–25% activity very high	Ajmaline, amitriptyline, carvedilol, codeine, flecainide, fluoxetine, galanthamine, haloperidol, metoprolol, mexiletine, ondansetron, propafenone, propranolol, tamoxifen, timolol, tropisetron
CYP3A4, CYP3A5, CYP3A7	Several mutations exist, some of them are rare and show reduced activity; expression of CYP3A7 in some adults only	Cyclosporin A, cortisol, dapsone, diltiazem, erythromycin, lidocaine, midazolam, nifedipine, paclitaxel, sildenafil, simvastatin, tacrolimus, triazolam, verapamil, zolpidem

[a] Frequencies for homozygous genotypes

Table 2. Hereditary polymorphisms in important enzymes metabolizing drugs and xenobiotics and their therapeutic consequences (Roots et al. 2004)

Other phase-I enzymes	Frequencies of genetic variants[a]	Examples of affected drugs
Flavin-dependent monooxygenase 3	Caucasians: 9% activity reduced	Albendazole, benzydamine, perazine, sulindac
Butyrylcholinesterase	Europeans: 0.03% activity lacking	Succinylcholine
Dihydropyrimidine dehydrogenase	Heterozygotes: 1% activity reduced	5-Fluoruracil
Phase-II enzymes		
Arylamine-N acetyltransferase 2 (NAT2)	Europeans: 55% slow acetylators	Dapsone, isoniazid, hydralazine, procainamid, sulfonamides
	Far East Asians: 17% slow acetylators	Dapsone, isoniazid, hydralazine, procainamid, sulfonamides
UDP-glucuronosyl-transferase 1A1	Caucasians: 10.9% activity reduced Asians: 1%–4% activity reduced	Irinotecan
Glutathione-S transferase GST M1	Europeans: 55% activity lacking	Risk of bladder cancer increased
Catechol-O methyltransferase	Caucasians: 25% activity reduced	Amphetamine, estrogen, L-dopa, α-methyldopa
Thiopurine-S methyltransferase	Caucasians: 0.3% activity lacking	Azathioprine, 6-mercaptopurine

[a] Frequencies for homozygous genotypes

ably higher activity is possible, as in the case of ultrafast metabolizers of cytochrome P450 2D6 (CYP2D6). Differences in activity might be less pronounced in heterozygous individuals than in homozygotes. Genetic variations (polymorphisms) in genes which encode drug-metabolizing enzymes primarily influence drug pharmacokinetics. However, genetic variations in drug transporters can also cause variable drug disposition.

6.2 Genotype–Phenotype Relations Leading to Dosage Recommendations

When a drug-metabolizing enzyme shows a clear phenotype–genotype correlation and the patient's genotype is known, the physician can adjust the dosage of drugs that are transformed by the particular enzyme genotype-specifically. About 7% of Caucasians are homozygously deficient in CYP2D6, i.e., they completely lack this enzymatic activity (Sachse et al. 1997) (Table 1). These individuals metabolize substrates of CYP2D6 considerably more slowly, with respective plasma levels being higher. The normal dose creates a supranormal response and those individuals experience more side effects caused by overdosage. Approximately 3% of Caucasians are carriers of a *CYP2D6* gene duplication. This is a very special case: these individuals have three active alleles that cause enzyme activity to be considerably higher than in wild-type carriers. Clinical studies indicate that a genotype-adapted dosage regimen could help avoid therapeutic failure and side effects.

Dosage regimens for some polymorphically metabolized drugs already exist (Brockmöller et al. 2000; Kirchheiner et al. 2001, 2004).

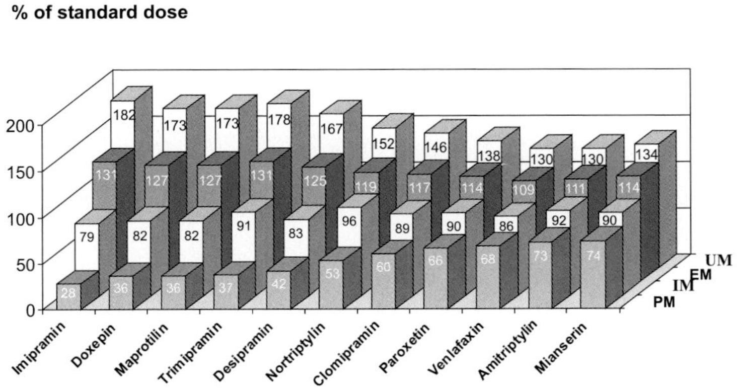

Fig. 2. CYP2D6 genotype-based dose recommendations for antidepressive agents. The calculations are based on a drug's recommended standard dosage. With the help of published genotype-specific pharmacokinetic data, dosages were calculated that will give similar plasma concentrations in all genotypes. *PM* poor metabolizer (CYP2D6 deficient), *IM* intermediate metabolizer (heterozygous wild-type with one deficient allele or with two alleles that lead to reduced activity), *EM* extensive metabolizer (homozygous wild-type), *UM* ultrafast metabolizer (one wild-type allele paired with one allele carrying a wild-type duplication) (Kirchheiner et al. 2001)

Figure 2 shows CYP2D6-based dose recommendations for antidepressive agents. Apart from polymorphic metabolic pathways, these calculations take into account the metabolite's pharmacodynamic potency. Genetic differences in enzymatic activity are without practical consequence if the metabolite and the mother substance are equally potent. But this is a rare case indeed, as most metabolites are without therapeutic effects.

Figure 3 illustrates the substantial differences in systemic clearance and the area under the concentration-time curve of the antidepressant doxepin (Kirchheiner et al. 2005). If the patients' CYP2D6 genotypes are known, individual dosage might equalize drug plasma concentrations.

With pro-drugs it is different. These pharmacologically inactive substances are transformed into active drugs in the body. Thus, only a small amount of codeine is O-demethylated to morphine by CYP2D6. Individuals who are homozygously deficient for CYP2D6 do not profit from morphine's cough relieving and analgesic effects because they do not form any morphine. The antimalarial proguanil is another example. CYP2C19 metabolizes proguanil to cycloguanil, the active antimalar-

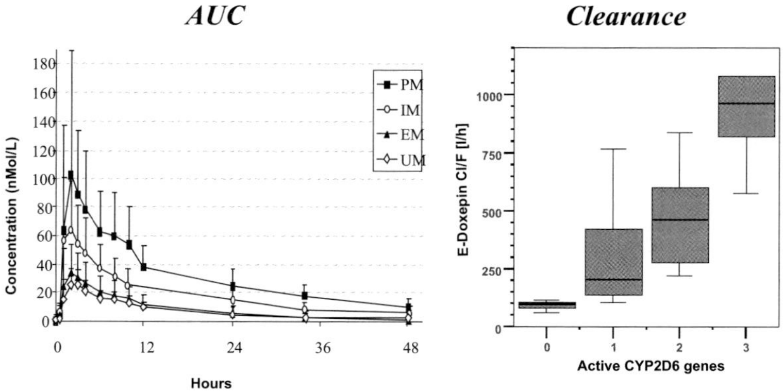

Fig. 3. CYP2D6 polymorphism and doxepin elimination. *Left panel* A single oral dose of 75 mg of racemic E-, Z-doxepin was given to healthy volunteers genotyped as extensive (*EM*), intermediate (*IM*), poor (*PM*), or ultrafast (*UM*) metabolizers. *Right panel* Oral clearance of the E-enantiomer of doxepin. (Kirchheiner et al. 2005)

ial principle. People with CYP2C19 deficiency transform proguanil via CYP3A4, but this pathway is less productive.

Polymorphisms in drug metabolizing enzymes do not exert the same effects on every drug: the pharmacokinetics of some of them are only slightly affected. Most drugs are metabolized by CYP3A4, which shows great interindividual difference in activity, although the genetic regulation has not yet been satisfactorily clarified (Table 1). Moreover, if there are alternative pathways that give polymorphic enzymes such as CYP2D6, CYP2C19, and CYP2C9 only a small share in a drug's biotransformation, their influence is negligible. Genotype-based dosing only makes sense if a clinically important polymorphism dominates the metabolism of a substance with small therapeutic range; these conditions are met by a rather small portion of our therapeutic equipment.

6.3 Genotype-Based Therapeutic Failure with 5-Hydroxytryptamine Type-3 Receptor Antagonists

Antiemetics are another example of how to optimize drug treatment. 5-Hydroxytryptamine type-3 receptor antagonists such as ondansetron and tropisetron have considerably improved antiemetic therapy in cancer patients. Nonetheless, 20%–30% of patients still suffer from emesis and nausea. 5-HT_3-antagonists are substrates of CYP2D6. Carriers of the gene for *CYP2D6* duplication were shown to be nonresponders to tropisetron and ondansetron (Fig. 4; Kaiser et al. 2002). These individuals are ultrafast metabolizers; obviously, they do not have drug plasma concentrations within the therapeutic range and should get considerably higher doses of tropisetron or ondansetron. There are only approximately 2%–3% ultrafast metabolizers in European Caucasian populations, but in Arabian and Northeast-African populations, the percentage rises to 5%–25% (see Table 1).

Apart from CYP2D6 polymorphism, antiemetic efficacy is also impaired by receptor variants. Tremblay et al. (2003) discovered that the AAG deletion in the promoter region of the 5-HT_3-receptor (-100 to -102 nucleotide position) was significantly more frequent in patients with unsatisfactory antiemetic response. Homozygous carriers of the deletion turned out to be nonresponders when given 5-HT_3-receptor an-

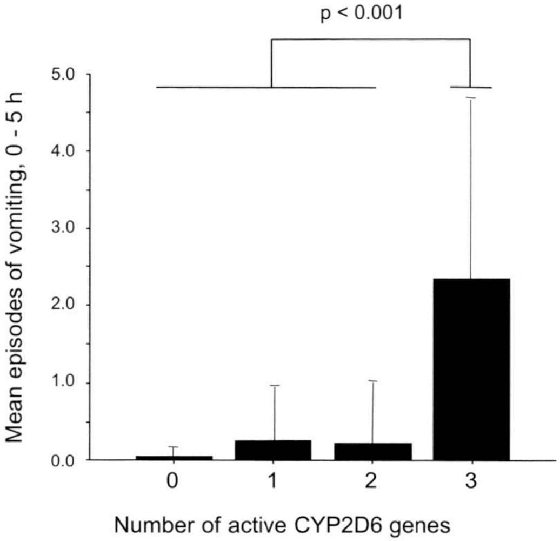

Fig. 4. Number of episodes of vomiting in relation to the *CYP2D6* genotype during the first 5 h after administration of emetogenic cancer chemotherapy. As illustrated, carriers of the *CYP2D6* gene duplication (genotype was mostly *CYP2D6**1/*2 × 2; i.e., three active genes) suffered from 2.3 episodes of vomiting on average, i.e., considerably more frequently than carriers of normal and deficient CYP2D6 metabolic phenotypes (carriers of 0, 1, and 2 active *CYP2D6* genes). (Kaiser et al. 2002)

tagonists. However, this genotype occurs only in about 1.5% of patients; therefore, its contribution to the total of therapeutic failures is only small.

6.4 Individual Variations in CYP2C19-Metabolism of Proton Pump Inhibitors Determine Their Efficacy

Proton pump inhibitors are very efficacious when used in the treatment of gastroesophageal reflux disease or *Helicobacter pylori* infection. Omeprazole is a good example to illustrate pharmacogenetics-based optimization of therapy. Roughly 80% of omeprazole is metabolized by CYP2C19 (Rost and Roots 1996). Individuals with a deficiency in CYP2C19 (poor metabolizers) transform omeprazole via CYP3A4 in a considerably slower reaction and show AUCs (area under the concentration-time curve) that are ten times larger than those of extensive metabolizers (Fig. 5).

Obviously, omeprazole is quite innocuous even if plasma concentrations are high. Therapeutic results seem to depend on the *CYP2C19*

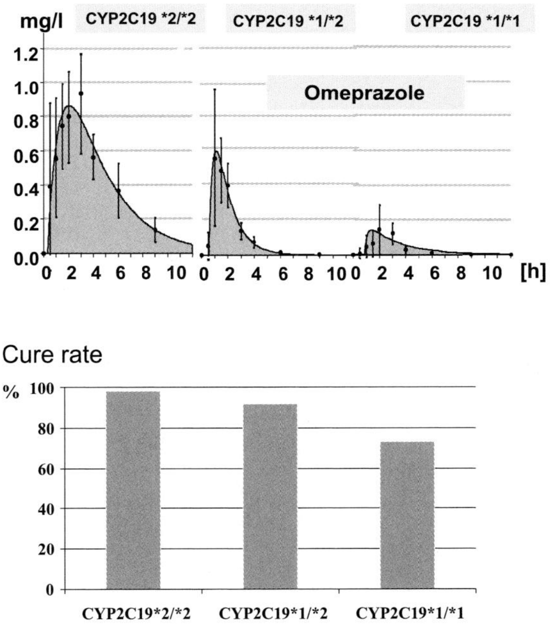

Fig. 5. *Upper panel* Plasma concentration time courses of 40 mg omeprazole p.o. in healthy volunteers with different *CYP2C19* genotypes. Individuals with the following genotypes were included (five per group): homozygous mutation (*CYP2C19**2/*2), phenotype: metabolism deficient because lacking CYP2C19 activity (poor metabolizer, PM); heterozygous wild-type/mutation (*CYP2C19**1/*2), phenotype: intermediate metabolizer (because of reduced metabolism); homozygous wild-type (*CYP2C19**1/*1), phenotype: extensive metabolizer, EM (Brockmöller et al. 2000). *Bottom panel Helicobacter pylori* cure rates in 62 peptic ulcer patients having received an eradication therapy. Patients are arranged according to their *CYP2C19* genotype (Furuta et al. 2001)

genotype. In several clinical studies, cure rates for *Helicobacter pylori* infection proved to be higher in homozygous carriers of the deficiency than in patients with two active genes (reviewed in Roots et al. 2004). The cure rates of heterozygous individuals with an active and an inactive gene ranged between these two groups. Most of these investigations were carried out in Japan and Korea where the deficiency in CYP2C19 occurs in 15%–25% of individuals, i.e., about five times more frequently than in white Europeans. If the physician knew the patient's *CYP2C19* genotype, it would be possible to take it into account when starting treatment and – to give an example – administer double the normal dose of omeprazole to homozygous wild-type carriers. However, there might be other factors that influence therapeutic results in addition to the CYP2C19 polymorphism, e.g., characteristics of resistance of *Helicobacter pylori* or polymorphisms in interleukin-1β (Take et al. 2003).

6.5 Cyclophosphamide Kinetics Related to CYP2C19 Polymorphism

Cyclophosphamide, a widely used cytostatic, is metabolized by a variety of polymorphic enzymes, particularly CYP enzymes. It is clinically well known that cyclophosphamide's anticancer efficacy and its tolerability show great interindividual variability. Cyclophosphamide is considered a pro-drug. In the body, it is metabolically activated by CYP2C9, CYP2B6, CYP2C19, and CYP3A to 4-hydroxycyclophosphamide. This primary metabolite furnishes the precursor material for the formation of alkylating agents, such as phosphoramide mustard.

We studied the plasma-elimination constant (k_e) of cyclophosphamide in a group of 49 cancer patients who received below 1,000 mg per m^2 of body surface, most of them together with other cytostatics. Figure 6 shows a clear gene–dose effect for the elimination constant when results were graded according to the *CYP2C19* genotype. The slowest elimination was observed – as expected – in poor metabolizers (genotype *CYP2C19*2/*2*; $k_e = 0.076$ [SD $= 0.014$] h^{-1}, mean). Elimination was 50% faster in individuals who were homozygous wild-type (*CYP2C19*1/*1*; $k_e = 0.113$ [SD $= 0.028$] h^{-1}). Future studies should

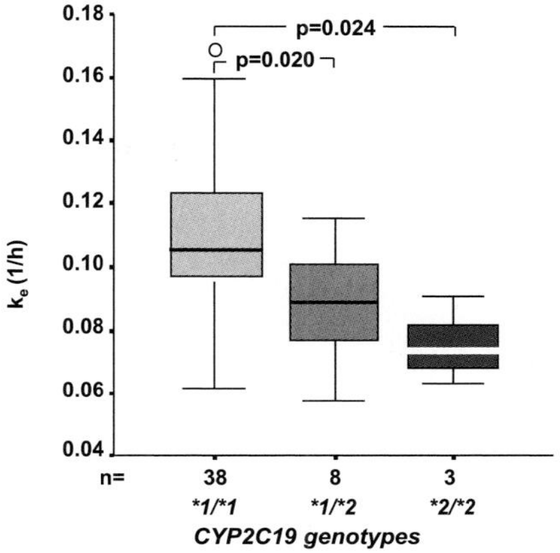

Fig. 6. Cyclophosphamide elimination rate (k_e) dependent on *CYP2C19* genotypes. Forty-nine cancer patients received cyclophosphamide doses below 1,000 mg/m^2. Results are given as box plots (Timm et al. 2005)

try to link these findings with clinical outcome. They should also include a pharmacogenetic evaluation of all the other enzymes involved in cyclophosphamide activation and disposition.

6.6 CYP2C9 Polymorphism

The CYP2C cluster on chromosome 10q24 does not only contain the gene encoding CYP2C19, but also genes encoding CYP2C9 and CYP2C8. The latter two are highly polymorphic enzymes that transform clinically important drugs, with some widely used antidiabetics among them (Table 1). The *CYP2C9*3* allele codes for a considerably reduced enzymatic activity. Figure 7 demonstrates how oral clearance of several antidiabetics is slowed down in *3-carriers. The greatest effect is seen with glyburide (glibenclamide). The homozygous genotype *CYP2C9*3/*3* occurs in about 0.4% of white subjects. Carriers of *CYP2C9*2/*2* (about 0.9% of white individuals) have a moderately reduced enzyme activity. In African and Far Eastern populations, these alleles are very rare. Whereas reduced CYP2C9 activity has been demon-

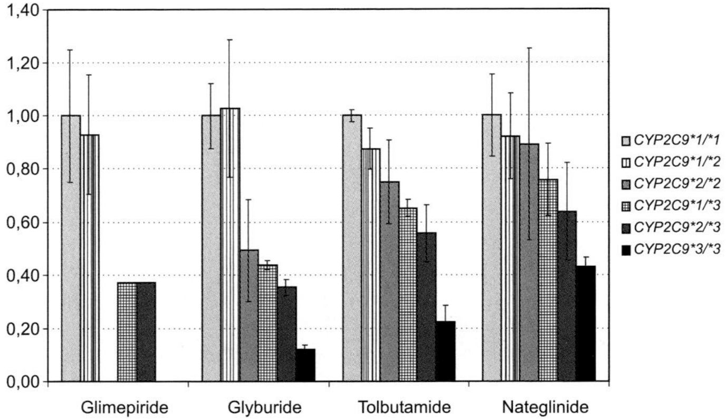

Fig. 7. Impact of CYP2C9 genetic variance on oral clearance of several antidiabetic drugs. Values observed for *CYP2C9** 1/* 1 (homozygous wild-type) were set to 1.0 (Kirchheiner et al. 2005, modified)

strated to be of major clinical consequence in the treatment with warfarin, its impact on antidiabetic therapy is still to be shown.

6.7 Estrogen Metabolism by CYP1A1 Variants

Figure 8 shows that endogenous substrates such as 17β-estradiol and estrone may also be transformed by polymorphic drug metabolizing enzymes (Kisselev et al. 2005). In this experiment, genotype–phenotype correlation was established for CYP1A1 variants that were expressed – together with P450 reductase – in *Spodoptera frugiperda* (Sf9) insect cells. In comparison to the wild-type protein, metabolite formation of the CYP1A1.2 enzyme variant is greatly enhanced, in particular the formation of 2-OH products. Due to the different pharmacodynamic profile of estrogen metabolites, these shifts in the metabolic patterns should also modulate susceptibility to estrogen-dependent diseases, such as osteoporosis, breast and ovarian cancer, and arteriosclerosis.

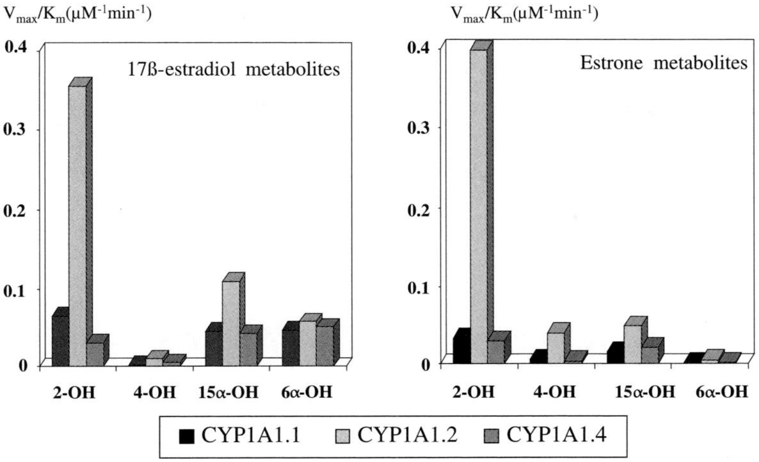

Fig. 8. Genetic variance in the metabolite patterns of 17β-estradial and estrone. CYP1A1 variants were expressed in Sf9 insect cells together with P450 reductase (Kisselev et al. 2005). Values of the catalytic efficiency are given (V_{max}/K_m)

6.8 Genotype–Phenotype Relations in Drug Transporters

Not long ago, drug tissue distribution was thought to be mainly a process of passive diffusion. Through the identification and characterization of a variety of transmembrane transporters, it has become clear that a considerable number of drugs are actively moved across biological membranes by transport mechanisms. Transmembrane transporters are integral membrane proteins and occur in several organs with absorptive and excretory functions (e.g., intestine, liver, and kidneys). Furthermore, they play an important role in forming blood–tissue barriers, such as the blood–brain barrier, the blood–placenta barrier, and the blood–testis barrier, thereby protecting these sensitive tissues against toxic xenobiotics. Transmembrane transporters can be subdivided into uptake and efflux carriers according to the principal direction of their substrate transfer into or out of the cell. Uptake carriers of pharmacokinetic interest include the transporter families OATP (*SLC21A*), OCT (*SLC22A*), and PEPT (peptide transporter; *SLC15A*). Efflux transporters of the ATPbinding cassette family (ABC), for instance P-glycoprotein, are major factors in the excretion of drugs.

Organic anion transporting polypeptides (OATPs) constitute a large family ubiquitously expressed in human tissues, such as liver, kidney, brain, and intestine. Liver OATPs serve in the extraction of xenobiotics and drugs from portal venous blood. Clinically relevant substrate drugs are, among others, the HMG-CoA reductase inhibitor pravastatin (Hsiang et al. 1999), the antihistamine fexofenadine, and the angiotensin-converting enzyme inhibitors enalapril and temocaprilat. OATP-C (*SLC21A6*) is exclusively found in the basolateral membranes of hepatocytes (Tamai et al. 2000). Recently, a variety of single nucleotide polymorphisms (SNPs) located within the exons of *OATP-C* has been discovered by different groups (Tirona et al. 2001; Nozawa et al. 2002; Michalski et al. 2002). Most of these variants are rare, with frequencies between 0.01 and 0.02. The most prevalent SNPs included A388G, C463A, and T521C, designated as *OATP-C*1b*, *4, and *5, respectively. Allele frequencies were significantly different in European-Americans, in African-Americans, and in Japanese subjects (Tirona et al. 2001; Nozawa et al. 2002). Most *OATP-C* SNPs were associated with altered in vitro transport characteristics of the transporter (Tirona et al. 2001).

Pharmacokinetic studies also showed that *OATP-C* haplotypes significantly affected the disposition of pravastatin (Mwinyi et al. 2004; Nishizato et al. 2003). Subjects bearing the *OATP-C*15* (Asp130Ala174) allele had reduced total and nonrenal clearances of pravastatin when compared to homozygous individuals of the *OATP-C*1b* genotype (Asp130Val174) (Nishizato et al. 2003). Significant pharmacogenetic effects of the *OATP-C* alleles *1*a*, *1*b*, and *5 on the 40-mg single-dose pharmacokinetics of pravastatin were also observed in a study that included healthy male whites from Germany (Fig. 9; Mwinyi et al. 2004). *OATP-C*5* expression appears to delay hepatocellular uptake of pravastatin, whereas *OATP-C*1b* seems to accelerate the OATP-C-dependent hepatic uptake of the drug.

Most efflux transporters involved in drug disposition belong to the ABC-family. P-glycoprotein (Pgp), the gene product of *MDR1* (*ABCB1*), is the pharmacogenetically best investigated efflux pump, followed by members of the MRP (ABCC) subfamily. Pgp was originally identified

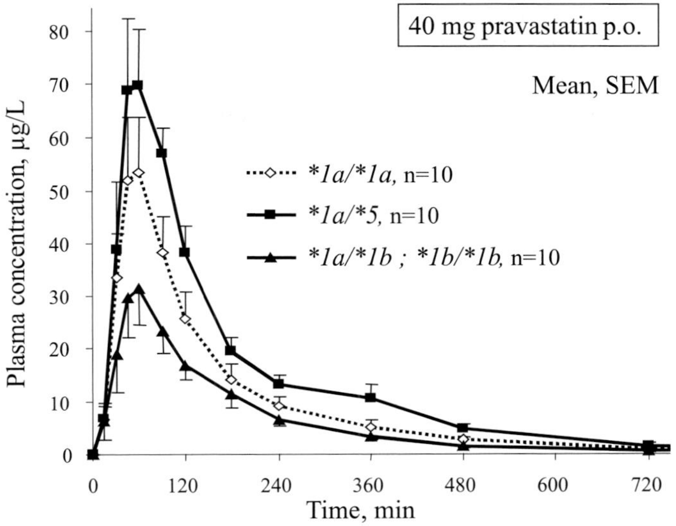

Fig. 9. Plasma concentration-time curves after a 40-mg oral dose of pravastatin in relation to tested *OATP-C* haplotypes in healthy volunteers. The Kruskal-Wallis test was performed for differences in AUC values across all three groups ($p = 0.006$). The Mann-Whitney test for comparison between the *1*a*/*1*a* and *1*a*/*5 group resulted in a p-value $= 0.049$ (Mwinyi et al. 2004)

as a major factor causing the multidrug resistance phenotype in tumor pharmacotherapy. However, Pgp's contribution to drug disposition has become evident due to its localization in excretory and absorptive tissues, including the canalicular (apical) membrane of hepatocytes, the brush border membrane of proximal tubular cells of the kidney, and the apical pole of enterocytes (Tanigawara 2000). In general, Pgp diminishes drug absorption and enhances drug excretion.

Polymorphic expression of Pgp was assumed to be an important modulator of individual response to pharmacotherapy. Indeed, several studies demonstrated pharmacogenetic effects of Pgp SNPs (Hoffmeyer et al. 2000; Hauser et al. 2005; Fellay et al. 2002; Johne et al. 2002). A noncoding SNP in exon 26 3435C→T, in particular, was associated with significantly lower intestinal Pgp expression levels, which led to higher steady-state plasma concentrations of digoxin in homozygous T-allele carriers in comparison to heterozygous or homozygous wild-type allele carriers (Hoffmeyer et al. 2000; Johne et al. 2002). However, there has been a large number of conflicting reports regarding the functional impact of this *MDR1* SNP (Johne et al. 2002; Sakaeda et al. 2001; Siegmund et al. 2002; Gerloff et al. 2002). The discrepancies in study results might partly be attributed to experimental conditions, for example steady-state vs single-dose kinetics, involvement of other transmembrane transporters, and the genetic surrounding of the *MDR1* locus. Interestingly, a haplotype-based analysis of two linked *MDR1* SNPs 2677G→T/A in exon 21 and 3435C→T in exon 26, could explain some of the contradictory results from previous studies (Johne et al. 2002). Thus, future investigations have to take into account the haplotype aspect.

6.9 Some Concluding Statements Regarding Pharmacogenetics

These examples illustrate that even today drug therapy could be tailored to the patient's individual genetic make-up. In accordance with the rules of evidence-based medicine, a pharmacogenetic dosage regimen should prove its superiority over standard methods in clinical studies. Preliminary results have already been presented for neuroleptics, antidepressive agents, azathioprine, and others.

Naturally, the demand for proofs by expensive clinical studies should not be pushed to an extreme. Inferences that are evident according to premium clinical and scientific knowledge do not require bureaucratic confirmation. For instance, the 50%-dose reduction of phenytoin in epileptics carrying the rare *CYP2C9* genotype *3/*3 can be deduced from existing knowledge (Brockmöller et al. 2000), and the efficacy of the individualized dose can be easily checked clinically. Haloperidol treatment of schizophrenic patients is a more difficult case, as outcome cannot be anticipated with certainty when pharmacotherapy is based on genotypes. Although haloperidol kinetics clearly depends on CYP2D6, an alternative, pharmacologically active metabolite is formed via carbonyl reduction. A clinical study, however, demonstrated the CYP2D6 dependence of haloperidol efficacy: poor metabolizers responded best, whereas ultra-fast metabolizers proved to be nonresponders (Brockmöller et al. 2002).

We consider pharmacogenetics as a part of future molecular medicine. Molecularly specified diagnosis will go hand in hand with molecularly specified drug therapy. We can expect that therapy will become safer. With fewer adverse side effects and the avoidance of ineffective therapy, even health care costs should decrease. The integration of pharmacogenetics into medical practice should be enhanced as soon as pharmacogenetic knowledge is integrated into the computer software used by physicians at the "point of drug prescription."

Acknowledgements. Part of the authors' research work was supported by the German Federal Ministry of Education and Research (BMBF), grant no. 03/4507 (InnoRegio Health Region Berlin-Buch "Pharmacogenomic optimization of drug therapy and drug development") given to CENiMED GmbH, Center for Individualized Medicine – Clinical Pharmacogenomics, Berlin.

References

Bönicke R, Lisboa BP (1957) Über die Erbbedingtheit der intraindividuellen Konstanz der Isoniazidausscheidung beim Menschen (Untersuchungen an eineiigen und zweieiigen Zwillingen). Naturwissenschaften 44:314

Brockmöller J, Kirchheiner J, Meisel C, Roots I (2000) Pharmacogenetic diagnostics of cytochrome P450 polymorphisms in clinical drug development and in drug treatment. Pharmacogenomics 1:125–151

Brockmöller J, Kirchheiner J, Schmider J, Walter S, Sachse C, Müller-Oerlinghausen B, Roots I (2002) The impact of the *CYP2D6* polymorphism on haloperidol pharmacokinetics and on the outcome of haloperidol treatment. Clin Pharmacol Ther 72:438–452

Fellay J, Marzolini C, Meaden ER, Back DJ, Buclin T, Chave JP, Decoster LA, Furrer H, Opravil M, Pantaleo G, Retelska D, Ruiz L, Schinkel AH, Vernazza P, Eap CB, Telenti A (2002) Swiss HIV Cohort Study. Response to antiretroviral treatment in HIV-1-infected individuals with allelic variants of the multidrug resistance transporter 1: a pharmacogenetics study. Lancet 359:30–36

Furuta T, Shirai N, Takashima M, Xiao F, Hanai H, Sugimura H, Ohashi K, Ishizaki T, Kaneko E (2001) Effect of genotypic differences in CYP2C19 on cure rates for Helicobacter pylori infection by triple therapy with proton pump inhibitor, amoxicillin, and clarithromycin. Clin Pharmacol Ther 69:158–168

Gerloff T, Schäfer M, Johne A, Oselin K, Meisel C, Cascorbi I, Roots I (2002) MDR1 genotypes do not influence the absorption of a single oral dose of 1 mg digoxin in healthy white males. Br J Clin Pharmacol 54:610–616

Hauser IA, Schaeffeler E, Gauer S, Scheuermann EH, Wegner B, Gossmann J, Ackermann H, Seidl C, Hocher B, Zanger UM, Geiger H, Eichelbaum M, Schwab M (2005) ABCB1 genotype of the donor but not of the recipient is a major risk factor for cyclosporine-related nephrotoxicity after renal transplantation. J Am Soc Nephrol 16:1501–1511

Hoffmeyer S, Burk O, von Richter O, Arnold HP, Brockmöller J, Johne A, Cascorbi I, Gerloff T, Roots I, Eichelbaum M, Brinkmann U (2000) Functional polymorphisms of the human multidrug-resistance gene: multiple sequence variations and correlation of one allele with P-glycoprotein expression and activity in vivo. Proc Natl Acad Sci U S A 97:3473–3478

Hsiang B, Zhu Y, Wang Z, Wu Y, Sasseville V, Yang WP, Kirchgessner TG (1999) A novel human hepatic organic anion transporting polypeptide (OATP2). Identification of a liver-specific human organic anion transporting polypeptide and identification of rat and human hydroxymethylglutaryl-CoA reductase inhibitor transporters. J Biol Chem 274:37161–37168

Johne A, Köpke K, Gerloff T, Mai I, Rietbrock S, Meisel C, Hoffmeyer S, Kerb R, Fromm MF, Brinkmann U, Eichelbaum M, Brockmöller J, Cascorbi I, Roots I (2002) Modulation of steady-state kinetics of digoxin by haplotypes of the P-glycoprotein MDR1 gene. Clin Pharmacol Ther 72:584–594

Kaiser R, Sezer O, Papies A, Bauer S, Schelenz C, Tremblay PB, Possinger K, Roots I, Brockmöller J (2002) Patient-tailored antiemetic treatment with 5-hydroxytryptamine type 3 receptor antagonists according to cytochrome P-450 2D6 genotypes. J Clin Oncol 20:2805–2811

Kalow W (1956) Familial incidence of low pseudocholinesterase level. Lancet 2:576–577

Kirchheiner J, Brøsen K, Dahl M, Gram L, Kasper S, Roots I, Sjöqvist F, Spina E, Brockmöller J (2001) CYP2D6 and CYP2C19 genotype-based dose recommendations for antidepressants: a first step towards subpopulation-specific dosages. Acta Psychiatr Scand Suppl 104:173–192

Kirchheiner J, Nickchen K, Bauer M, Wong ML, Licinio J, Roots I, Brockmöller J (2004) Pharmacogenetics of antidepressants and antipsychotics: the contribution of allelic variations to the phenotype of drug response. Mol Psychiatry 9:442–473

Kirchheiner J, Roots I, Goldammer M, Rosenkranz B, Brockmöller J (2005) Effect of genetic polymorphisms in cytochrome P450 (CYP) 2C9 and CYP2C8 on the pharmacokinetics of oral antidiabetic drugs: clinical relevance. Clin Pharmacokinet 44:1209–1225

Kisselev P, Schunck WH, Roots I, Schwarz D (2005) Association of CYP1A1 polymorphisms with differential metabolic activation of 17beta-estradiol and estrone. Cancer Res 65:2972–2978

Michalski C, Cui Y, Nies AT, Nuessler AK, Neuhaus P, Zanger UM, Klein K, Eichelbaum M, Keppler D, Konig J (2002) A naturally occurring mutation in the SLC21A6 gene causing impaired membrane localization of the hepatocyte uptake transporter. J Biol Chem 277:43058–43063

Mwinyi J, Johne A, Bauer S, Roots I, Gerloff T (2004) Evidence for inverse effects of *OATP-C (SLC21A6)* *5 and *1*b* haplotypes on pravastatin kinetics. Clin Pharmacol Ther 75:415–421

Nishizato Y, Ieiri I, Suzuki H, Kimura M, Kawabata K, Hirota T, Tarane H, Irie S, Kusuhara H, Urasaki Y, Urae A, Higuchi S, Otsubo K, Sugiyama Y (2003) Polymorphisms of OATP-C (SLC21A6) and OAT3 (SLC22A8) genes: consequences for pravastatin pharmacokinetics. Clin Pharmacol Ther 73:554–565

Nozawa T, Nakajima M, Tamai I, Noda K, Nezu J, Sai Y, Tsuji A, Yokoi T (2002) Genetic polymorphisms of human organic anion transporters OATP-C (SLC21A6) and OATP-B (SLC21A9): allelic frequencies in the Japanese population and functional analysis. J Pharmacol Exp Ther 302:804–813

Roots I, Gerloff T, Meisel C, Kirchheiner J, Goldammer M, Kaiser R, Laschinski G, Brockmöller J, Cascorbi I, Kleeberg U, Hildebrandt AG (2004) Pharmacogenetics-based new therapeutic concepts. Drug Metab Rev 36:617–638

Rost KL, Roots I (1996) Nonlinear kinetics after high-dose omeprazole caused by saturation of genetically variable CYP2C19. Hepatology 23:1491–1497

Sachse C, Brockmöller J, Bauer S, Roots I (1997) Cytochrome P450 2D6 variants in a Caucasian population: allele frequencies and phenotypic consequences. Am J Hum Genet 60:265–271

Sakaeda T, Nakamura T, Horinouchi M, Kakumoto M, Ohmoto N, Sakai T, Morita Y, Tamura T, Aoyama N, Hirai M, Kasuga M, Okumura K (2001) MDR1 genotype-related pharmacokinetics of digoxin after single oral administration in healthy Japanese subjects. Pharm Res 18:1400–1404

Siegmund W, Ludwig K, Giessmann T, Dazert P, Schroeder F, Sperker B, Warzok R, Kroemer HK, Cascorbi I (2002) The effects of the human MDR1 genotype on the expression of duodenal P-glycoprotein and disposition of the probe drug talinolol. Clin Pharmacol Ther 72:572–583

Take S, Mizuno M, Ishiki K, Nagahara Y, Yoshida T, Inaba T, Yamamoto K, Okada H, Yokota K, Oguma K, Shiratori Y (2003) Interleukin-1β genetic polymorphism influences the effect of cytochrome P2C19 genotype on the cure rate of 1-week triple therapy for Helicobacter pylori infection. Am J Gastroenterol 98:2403–2408

Tamai I, Nezu J, Uchino H, Sai Y, Oku A, Shimane M, Tsuji A (2000) Molecular identification and characterization of novel members of the human organic anion transporter (OATP) family. Biochem Biophys Res Commun 273:251–260

Tanigawara Y (2000) Role of P-glycoprotein in drug disposition. Ther Drug Monit 22:137–140

Timm R, Kaiser R, Lötsch J, Heider U, Sezer O, Weisz K, Montemurro M, Roots I, Cascorbi I (2005) Association of cyclophosphamide pharmacokinetics to polymorphic cytochrome P450 2C19. Pharmacogenomics J 5:365–373

Tirona RG, Leake BF, Merino G, Kim RB (2001) Polymorphisms in OATP-C: identification of multiple allelic variants associated with altered transport activity among European- and African-Americans. J Biol Chem 276:35669–35675

Tremblay PB, Kaiser R, Sezer O, Rosler N, Schelenz C, Possinger K, Roots I, Brockmöller J (2003) Variations in the 5-hydroxytryptamine type 3B receptor gene as predictors of the efficacy of antiemetic treatment in cancer patients. J Clin Oncol 21:2147–2155

7 Clinical Trials in Elderly Patients

S.H.D. Jackson

Abstract. The increasing size of the elderly population means that both the relative and absolute numbers of prescriptions for elderly patients are increasing. Depending on the age group, between 60% and 80% of elderly people are taking medication, and between 20% and 30% are taking at least three drugs. Prescribing for elderly patients as opposed to younger patients is thus ever more important. This has inevitably meant that the drug development process must increasingly recognise the importance of identifying and developing therapeutic targets relevant to older patients. Clearly, the scientific ethical and regulatory principles that determine conduct of clinical trials in younger individuals apply equally to older people. In addition, the development of drugs to be used in older patients requires an awareness of a number of physiological, pathophysiological and sociological considerations.

7.1 Pharmacokinetics

Pharmacokinetics can be considered as the way in which the body handles drugs (Table 1). Age, however, is only one of many factors affecting pharmacokinetics. A number of age-related changes in physiology give rise to changes in pharmacokinetics associated with ageing (Table 2). There is no convincing evidence that age affects the rate or extent of absorption (Gainsborough et al. 1993).

Table 1. Pharmacokinetics vs pharmacodynamics

Pharmacokinetics
Liberation
Absorption
Distribution
Metabolism
Elimination
Pharmacokinetic interactions
(Drug A alters the pharmacokinetics of drug B)
Pharmacodynamics
Pharmacodynamic interactions
(Drug A alters the pharmacodynamics of drug B independently of pharmacokinetic changes)

Table 2. Age-related changes relevant to pharmacokinetics

Reduced glomerular filtration (↓ CL for water-soluble drugs)
Reduced liver volume (↓ CL for lipid-soluble drugs)
Increased ratio of body fat to water (↑ V hence ↑ $t^1/_{2z}$ for lipid-soluble drugs)
Reduced hepatic enzyme activity (in frail elderly patients) (↓ CL for lipid-soluble drugs)
Reduced protein binding (in frail elderly patients) (↑ V hence ↑ $t^1/_{2z}$)

7.1.1 Reduced Glomerular Filtration

The glomerular filtration rate tends to decrease with age, and this affects elimination of drugs cleared renally (water-soluble drugs). In some individuals, glomerular filtration may fall by more than 50% during the second half of their lives. Thus, the elimination half-life of drugs cleared wholly by renal excretion may double. Where the toxic level is close to the therapeutic level (a narrow therapeutic window), such a change can lead to toxic effects.

Digoxin is probably the most widely used drug in this category. While younger individuals typically require 0.25 mg digoxin as a maintenance dose, patients over the age of 70 years usually require half this dose, or even less if renal disease is present.

Serum creatinine rises above the normal range only when a very substantial fall in the glomerular filtration rate has occurred. Serum creatinine is also determined by muscle mass. Thus serum creatinine is not a good marker of renal function. The most useful surrogate marker of clearance for renally cleared drugs is creatinine clearance. This is most easily derived using the well-known Cockroft–Gault equation (Cockroft and Gault 1976).

7.1.2 Reduced Tubular Secretion

For those few drugs that have been studied, renal tubular secretion is also impaired by ageing, but this is unlikely to be clinically significant without a parallel change in glomerular filtration.

7.1.3 Reduced Hepatic Clearance

Lipid-soluble drugs are predominantly cleared by metabolism in the liver. For most drugs cleared in this way, the hepatic extraction ratio (the proportion of molecules removed from the blood during passage through the liver) is low and mainly determined by liver volume. Liver volume falls by about 30% during adult life. Therefore, as a result of normal ageing, there is a reduction in hepatic clearance of such drugs due to a reduction in liver volume (Wynne et al. 1989).

The variability in the decline of hepatic clearance with age is much greater than that of renal clearance. As a result, there is considerable

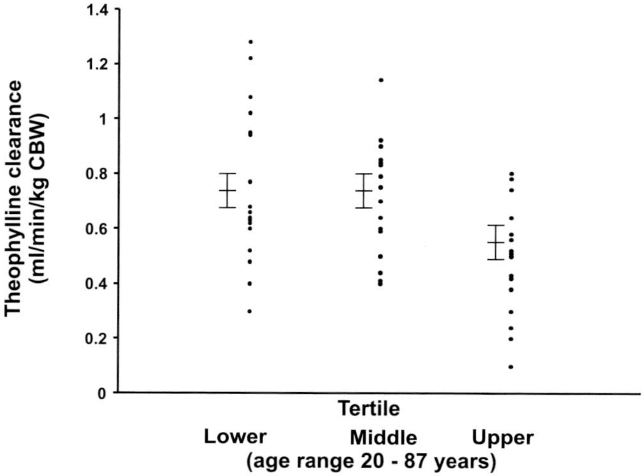

Fig. 1. Age associated change in theophyllinic clearance expressed as mls/min/kg corrected body weight (CBW) (85% theophylline clearance is via hepatic metabolism)

overlap in clearance values (and hence required dosage) between young and elderly individuals (Fig. 1) (Jackson et al. 1989).

7.1.4 Increased Proportion of Body Fat

Normal ageing is associated with a reduction in the relative ratio of body water: body fat. The most important pharmacokinetic consequence of this is the increase in volume of distribution of lipid-soluble drugs such as benzodiazepines. This prolongs the half-life according to the relationship:

$$t^1/_{2z} \propto \frac{V}{CL}$$

where

$$t^1/_{2z} = \text{elimination half-life}$$
$$V = \text{volume of distribution}$$
$$CL = \text{clearance}$$

Thus the effect of ageing on the volume of distribution of benzodiazepines is to prolong $t^1/_{2z}$. There will, in addition, be an effect of reduced liver volume on clearance. This also results in a prolongation of $t^1/_{2z}$ independently of an effect via V.

7.1.5 Frailty

Certain disease processes, in combination with ageing, may lead to changes referred to as frailty. Although this is difficult to define, it may be associated with some form of dependence and manifestations of nonspecific ill health, e.g. a normochromic normocytic anaemia and hypoalbuminaemia. These patients have reduced hepatic enzyme activity, which further reduces the clearance of hepatically metabolised drugs. In addition, the reduced serum albumin results in reduced protein binding of acidic drugs. For heavily protein-bound drugs such as ibuprofen (99.7% protein bound), this reduction can result in an increase in the volume of distribution and hence the half-life. This happens because there is less protein to "hold" the drug in the plasma, resulting in diffusion into the tissues.

7.2 Pharmacodynamics

Pharmacodynamic effects of ageing have not been studied as extensively as pharmacokinetics, although changes associated with ageing have been demonstrated for some drugs. For example, elderly patients may have increased sensitivity to drugs such as warfarin and benzodiazepines (Table 3) (Castleden et al. 1977; Shepherd et al. 1977). Sensitivity is defined as the response to a given concentration of drug. Sensitivity is independent of age-related changes in the pharmacokinetics of drugs. Thus, for example, the increased effect of digoxin in an elderly patient relates to higher concentrations (as a result of reduced renal clearance), rather than increased sensitivity. However, the increased anticoagulant effect of warfarin is at least partly a result of increased sensitivity to the drug, rather than higher concentrations alone.

Increased sensitivity often reflects an underlying loss of functional reserve, for example, the gastrointestinal tract and NSAIDs, cognitive function and both anticholinergics and benzodiazepines, posture and

Table 3. Drugs for which changes in pharmacodynamics are associated with ageing. Changes in sensitivity are indicated as increased (↑) or decreased (↓)

Benzodiazepines (↑)
Warfarin (↑)
Hypotensives (↑)
Calcium channel-blocking effects on PR interval (↓)
Anticholinergics (↑)
Beta-adrenergic modulators in some tissues (↓)
Nonsteroidal anti-inflammatory drugs (↑ gastrointestinal adverse reactions)
Phenothiazines (↑)

benzodiazepines, blood pressure homeostasis and hypotensives, and thermoregulation and phenothiazines.

7.3 Adverse Drug Reactions

There are two types of adverse drug reactions (ADRs): dose-dependent and idiosyncratic. In addition, a group of reactions has been described as pseudo-allergic. They are usually idiosyncratic reactions (for example, the chlorpropamide alcohol flush), which mimic allergic reactions because substances such as histamine are released. There is no evidence that they occur more commonly in elderly patients. Most of the excess in older patients is due to dose-related ADRs.

The rise in ADRs with increasing age is multifactorial. Changes in both pharmacokinetics and pharmacodynamics associated with ageing account for part of the excess, but the increased prevalence of disease states, together with the consequent medication, also contribute.

Digoxin is a well-known cause of ADRs in elderly patients because of reduced renal clearance. On the other hand, the increased incidence of ADRs to diuretics is caused predominantly by increased sensitivity to the effects of diuresis (as opposed to increased sensitivity to diuretics themselves). Adverse reactions to benzodiazepines are also the result of both direct effects of the drug and secondary effects. The incidence of gastrointestinal ADRs to NSAIDs, corrected for the number of prescriptions is similar in young and elderly patients. However, the severity of the reactions among elderly patients is greater, and life-threatening perforations and haemorrhage are more common.

7.4 Drug Interactions

As for ADRs, the increased need for drug treatment with increasing age, together with the changes in pharmacokinetics and pharmacodynamics, increases the risk of clinically significant drug interactions.

There are two types of drug interactions: pharmacokinetic interactions, where drug A changes the pharmacokinetics of drug B, and pharmacodynamic interactions where drug A changes the pharmacodynamics of drug B without changing the pharmacokinetics.

As in younger patients, interactions involving warfarin, lithium and digoxin have great potential to produce serious clinical effects because the toxic levels are close to the therapeutic levels (narrow therapeutic window).

The choice of what interaction studies to undertake during clinical drug development will be determined by the lipid/water solubility of the drug and the likely drugs to be co-administered. Thus, for lipid soluble drugs interactions with other drugs metabolised by the same hepatic enzyme will need to be investigated, e.g. SSRIs inhibit CYP450 2D6.

7.5 Clinical Trial Design

In addition to addressing the relevance of age-related changes in pharmacokinetics and pharmacodynamics during the clinical development of a drug, there are a number of important considerations related to undertaking both healthy volunteer studies and clinical trials in older patients.

7.5.1 Motivation

Our experience is that only 6% of healthy elderly volunteers take part in studies for the volunteer fees, whereas the figure for young volunteers is around 95% (unpublished observations). The majority of healthy elderly volunteers cite their wish to do something worthwhile, to help others or the scientific interest. Interestingly, 75% see volunteering as a way of saying thank you for past therapeutic benefit.

7.5.2 Time Constraints

The perception that without the constraint of work, retired people have a greater ability to conform to clinical schedules than younger people is mistaken. This is particularly true for elderly volunteers who tend to be from higher socioeconomic groups than elderly patients taking part in clinical trials. These individuals often have very busy schedules and spend time away from home.

7.5.3 Incarceration

Particularly for early clinical development studies, it is usual to plan to incarcerate subjects to improve efficiency. Unlike healthy young volunteers, this is a major barrier to recruitment and cooperation cannot be ensured by increasing the volunteer fees.

7.5.4 Exclusion Criteria

Exclusion criteria appropriate for younger patients and volunteers need careful scrutiny before being applied to older subjects. This is because the prevalence of minor ailments rises with age. In association with this, the use of analgesics and over-the-counter drugs increases. Whilst drugs known or suspected to interact with investigational drugs must always be excluded, many other agents have no theoretical reasons to be exclusion criteria, yet often are.

7.6 Conclusions

Healthy elderly volunteer studies and clinical trials in elderly patients must be undertaken when an investigational drug will be used in an elderly population. These studies must be designed to identify age-related changes in pharmacokinetics and pharmacodynamics, as well as clinically relevant drug interactions. The design of such studies in older people needs to take account of the changes in circumstances and the prevalence of minor ailments and therapeutic drug consumption.

References

Castleden CM, George CF, Mercer C, Hallet C (1977) Increased sensitivity to nitrazepam in old age. Brit Med J 1:10–12

Cockroft DW, Gault MH (1976) Prediction of creatinine clearance from serum creatinine. Nephron 16:31–41

Gainsborough N, Maskrey VL, Nelson ML, Keating J, Sherwood RA, Jackson SHD, Swift CG (1993) The association of age with gastric emptying. Age Ageing 22:37–40

Jackson SHD, Johnston A, Woolard R, Turner P (1989) The relationship between theophylline clearance and age in adult life. Eur J Clin Pharmac 36:29–34

Shepherd AM, Hewick DS, Moreland TA, Stevenson IH (1977) Age as a determinant of sensitivity to warfarin. Br J Clin Pharmacol 4:315–320

Wynne H, Cope LH, Mutch E, Rawlins MD, Woodhouse KW, James OFW (1989) The effect of age upon liver volume and apparent liver blood flow in healthy man. Hepatology 9:297–301

8 Dose Finding in Pediatric Patients

G. Henze

Abstract. It is generally agreed that satisfactory safety and effectiveness of pharmaceutical products for children and adolescents have not yet been established. This applies in particular to anti-cancer drugs and even to those having successfully been used for many years in multidrug chemotherapy protocols for childhood cancer. For example, nephroblastoma or Wilms' tumor is one of the typical and frequent forms of childhood cancer occurring at a median age of about 3 years. Standard therapy for Wilms' tumor is the combination of vincristine and actinomycin D; survival is about 85%. For actinomycin D, the summary of product characteristics states that one contraindication is children aged below 6–12 months. If this would be considered and respected it would mean that a substantial proportion of children with Wilms' tumor would not be treated and thus a proven curative therapy would be withheld. The current situation in pediatrics is that off-label use has become a common practice: in private practice about 20% of prescriptions are off-label, in children's hospitals approximately 40%–50% with 50%–70% in pediatric oncology and more than

90% in neonatology (Conroy et al. 1999, 2000, 2003; Turner et al. 1996, 1998; McIntyre et al. 2000). These conditions are more or less tolerated by the authorities although they are beyond legality. The reason is that appropriate clinical trials like those in adults have not been conducted in children and drugs have therefore not been licensed.

8.1 Peculiarities of Childhood and Adolescence

Childhood and adolescence are characterized by different phases of maturation and differentiation of organs, the immune system, and the central nervous system (Brochhausen and Seyberth 2000). The most immature individuals are preterms with their specific problems such as surfactant deficiency, persistent fetal circulation, immaturity of the brainstem, lack of autoregulation of circulation, and incomplete vascularization of the retina, which result in problems of pulmonary adaptation, pulmonary hypertension, persistent ductus arteriosus, bronchopulmonary dysplasia, and retinopathy of prematurity. Newborns have a large body surface area and a high content of body water and fat. In infants, the immune system is still incompetent, the myelinization of nerves incomplete, and the maturation of metabolic processes still in progress. School children have a lower growth rate and are becoming increasingly independent. Finally, in adolescents puberty starts with again a higher growth rate, and emotional instability is common.

Hence, children are not simply small adults, and for pharmacology and drug dosing this means that drug doses cannot merely be converted from adults according to body weight or body surface area.

8.2 Reasons for the Lack of Clinical Trials in Children

Children, and in particular research in children, are quite strictly protected by law. Thus, the general conditions for clinical trials are much more complicated than in adults because the legal conditions frequently hamper pharmaceutical studies, in particular those that have no direct implications for the individual child. Frequently, ethical reasons do not allow clinical and/or pharmaceutical studies in children, and there are ethics committees that require the consent of the patient and not just the consent of the parents, in particular if the intended trial involves more than "minimal suffering" of the child, a term that needs to be clearly defined. Further problems involving conducting clinical trials in children are low patient recruitment meaning that frequently multicenter trials have to be carried out in order to get results in a reasonable time span. Often children's hospitals do not have sufficiently trained staff for clinical trials. Finally, there is a low market potential for pharmaceutical drugs in children, rendering clinical trials and drug development economically unattractive to the pharmaceutical industry.

8.3 How to Select the Appropriate Dose?

Many drugs used in pediatric practice have not been studied adequately or at all in children. Many formulas have been suggested to calculate the pediatric dose from the adult dose (e.g., Clark's, Cowling's, and Young's rules), assuming *incorrectly* that the adult dose is always right and that the child is a miniature adult. However, based on the particularities mentioned above the dosage requirements constantly change with age and maturation (Merck 1999).

In general, drug dosing is calculated according to the body surface area (BSA). The BSA corresponds well to the extracellular fluid and to metabolic parameters such as cardiac and renal function. In addition, drug concentrations can easily be measured in serum, which is part of the extracellular space. However, in newborns and infants the extracellular space is much higher than in older children and adults in relation to body weight. An example is shown in Fig. 1. Most often, a body weight of 30 kg is set to the equivalent of 1 m^2 of body surface area. This is a relatively rough but realistic estimation. If this ratio is applied to all ages during childhood it can easily be seen that very young children would receive much higher doses (in newborns about 60% more) than older children. In contrast, in children over the age of 10 years the doses would be lower, over the age of 15 years up to about 20%. Without appropriate pharmacokinetic studies, it is impossible to determine the correct dosage; and even if serum levels are measured one cannot be sure that the measured concentration will have the correct biological effect.

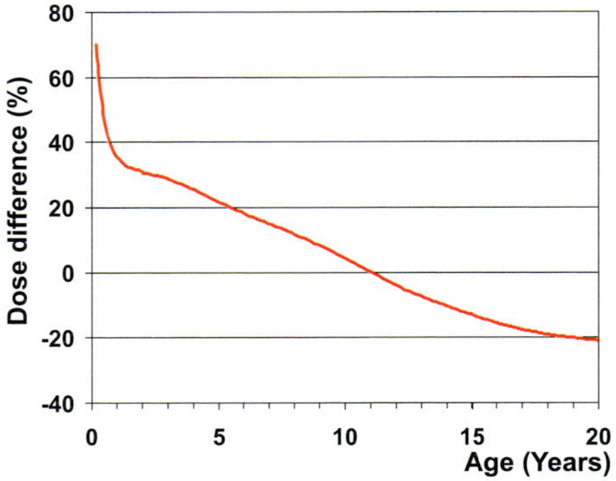

Fig. 1. Percentage dose difference between body surface area and body weight based dosing assuming 30 kg BW = 1 m^2 BSA

Table 1. Recommendations for dose adjustments according to age in acute lymphoblastic leukaemia (Protocol ALL BFM 2000, courtesy of M. Schrappe)

All drugs	Dose reduction	Basis
$\leq$ 6 months	1/3	BSA
7–12 months	1/4	BSA
$\geq$ 12 months	None	BSA

In order to avoid overdosing in very young children (mostly aged less than 1 year) various rules have been established to adjust the BSA-based dose to the young. Examples of these procedures are shown in Table 1. Frequently, the drug dose is adjusted according to the following procedure:

$$1\ \mathrm{m}^2 \mathrel{\hat{=}} 30\ \mathrm{kg}$$

Consequently the BSA dose can be adjusted according to:

$$\mathrm{mg/m^2} = x/30\ \mathrm{mg/kg}$$

For example, a scheduled dose of 1,000 mg/m^2 of cyclophosphamide for an infant with a body weight of 10 kg and a height of 75 cm were calculated as 330 mg following this rule. If the same child was 13 months of age, i.e., just beyond infancy, the calculated BSA would be 0.46 m^2 and the corresponding cyclophosphamide dose would thus be 460 mg.

If body weight were used to calculate the dose, another example might illustrate the problems. VP 16 is used in conditioning regimens prior to stem cell transplantation, usually at a dose of 60 mg/kg. For a 15-year-old patient with a body weight of 60 kg this would result in a dose of 3,600 mg. If this patient were obese with a body weight of 120 kg the dose would be calculated at 7,200 mg, a dose which is likely to be lethal.

Again a different regimen to adjust drug doses for young children is being used in the protocol for treatment of acute lymphoblastic leukemia of the Berlin-Frankfurt-Münster group ALL BFM 2000 (Table 1).

All of the examples mentioned show that there is no reliable procedure to define the appropriate dose of drugs in children and that the currently used regimens are often rough estimations. They reflect insufficient knowledge and helplessness. In part they even deny biological principles because steps of 20%, 30% or more at a certain cut-off age are absolutely unrealistic.

8.4 Do We Need New Drugs?

The urgent need for new drugs has often been emphasized in public in light of improved but still insufficient cure rates, in particular in childhood cancer. However, the examples mentioned clearly demonstrate that we still do not know how to use the "old" drugs properly. Only recently, the US Children's Cancer Group published data showing that in acute lymphoblastic leukemia, dexamethasone is superior to prednisone (Bostrum et al. 2003). The 6-year event-free survival by randomized treatment was $85\% \pm 2\%$ for dexamethasone vs $77\% \pm 2\%$ for the prednisone group ($= 0.002$). Similarly, the 6-year risk of isolated central nervous system relapse by randomized glucocorticoid was $3.7\% \pm 0.8\%$, significantly lower in the DEXA than in the PRED group ($7.1\% \pm 1.1\%$; $p = 0.01$).

Marked differences were found between the biological activity of different preparations of L-asparaginase (Vieira Pinheiro et al. 1999; Boos et al. 1996). These data are in agreement with the results of an EORTC clinical trial for childhood acute lymphoblastic leukemia (Duval et al. 2002). In the group of children who were treated with *Erwinia chrysan-*

temi asparaginase, the remission failure rate was 4.9% contrasting with only 2.0% in children having received *Escherichia coli* asparaginase ($p = 0.038$). The 6-year event-free survival by randomized treatment was $59.8 \pm 2.6\%$ in the *Erwinia* vs $73.4 \pm 2.4\%$ in the *E. coli* group ($p = 0.0004$), and the 6-year overall survival was $75.1 \pm 2.3\%$ vs $83.9 \pm 2.0\%$ ($p = 0.002$).

Concerning asparaginase, for example, even recent publications have shown that we still do not know the optimum dose of the drug. Likewise we do not know the duration of required asparagine depletion. It was shown, however, that different asparaginase preparations have different biological and clinical activities and it appears to be clear that *Erwinia* asparaginase cannot be substituted for native *E. coli* asparaginase. It is not yet clear whether the pegylated product PEG asparaginase can be substituted for native *E. coli* asparaginase, in particular because it is not clear whether PEG asparaginase is active in the central nervous system. We do not know exactly which factors are responsible for hyperreactivity against asparaginase; the same is true for silent inactivation of the drug. Finally, we cannot properly assess the hazards when asparaginase is given together with glucocorticoids, in particular in patients with thrombophilic states.

Thus, in order to answer all of the questions raised concerning a single drug, we obviously need clinical trials in children as urgently as we may need new drugs. Such clinical trials must meet ethical and quality requirements as they must meet the therapeutic needs of children in order to avoid unnecessary studies. It is essential that all information on ongoing trials and results be available to avoid duplication of studies. A key role for such trials, especially in Europe with the recent Clinical Trials Directive, should be given to the Pediatric Committee of the European Medicine Evaluation Agency (EMEA). As in the US, more attention should be given to pediatric initiatives where with incentives, rewards, and obligations, a substantial increase in pediatric studies, labels, and new anti-cancer drugs could be achieved.

In the EU, a network for clinical trials should be established through the International Society of Paediatric Oncology (SIOP)-Europe and national groups. Major concerns in this respect are currently costs and funding. A European consortium for prioritization and early evaluation of new compounds (Innovative Therapies for Children with Cancer,

ITCC) has been established. They are acting in close collaboration with groups in the US. In Europe, there is a partnership with parents and patients through SIOP, and a collaboration with the EMEA and the FDA has been started in order to provide guidelines and favor international studies. Some pharmaceutical companies have expressed their interest in participating in such developments.

8.5 The Role of Therapy Optimizing Studies in Pediatric Oncology

Therapy optimizing studies (TOSs) have been major tools for the substantial progress in the treatment of childhood cancer during the past 30–40 years (Creutzig et al. 2003, 2005). The principle of TOS is as simple as it is effective: treatment is given according to a fixed protocol in the form of a multi-centric cooperative, mostly randomized trial. The results are carefully evaluated and form the basis of the consecutive trial, aiming at eliminating weaknesses of preceding therapies. In pediatric oncology, TOSs are necessary for several reasons. They test a necessary, multimodal therapy with curative intention according to current scientific standards. Patients are treated according to stratified, risk-adapted protocols that frequently contain a randomized question. Because of the low patient numbers, they are multi-centric. TOSs are subject to drug legislation but at the same time they are tools for treatment, clinical research, and quality assurance. In contrast to mere pharmaceutical/pharmacological trials, TOSs are not aimed at licensing drugs.

Between 1980 and 2003 in Germany, 35,367 children under the age of 15 years were registered in the German Childhood Cancer Registry in Mainz (Anonymous 2006). About 92% of these children were treated in TOSs. The high participation rate is unique in the world and has led to excellent results in the survival of childhood cancer. For the most common form of childhood malignancies, acute lymphoblastic leukemia, the overall survival is approaching 90%. Interestingly and remarkably, the drugs that are being used for treatment are almost essentially the same as those used in the 1970s. Not a single new drug has been responsible for the steadily improved treatment results. The most important steps for progress were changes in dosing and timing of known drugs and in general intensification of therapy with the appropriate use of stem cell transplantation for the patients in whom chemotherapy alone was not likely to be curative.

Thus, TOSs have proven to be highly effective tools for the treatment of childhood cancer and the improved and impressive results have been achieved using "old" drugs. For treatment in general, the pharmacological basis is often insufficient, and it would be worthwhile to reconsider the therapeutic targets and the targeted compartments, and also to define the distribution volumes of drugs, i.e., the extracellular space or the intracellular space. For the future it appears of utmost importance to aim at the development of "intelligent" drugs, i.e., those with a more specific anti-cancer effect and less acute and late adverse effects.

8.6 The Future

In recent years, increasing information has become available on the role of genetics in cancer. This does not only mean genetic alterations of cancer cells as compared with normal cells. Genetics is also important in order to assess the effects and side effects of certain drugs in individual patients. Polymorphisms of drug-metabolizing enzymes have been found and described, which can in part explain differential effects in different individuals. At a given dose, drugs may be ineffective in some while causing severe side effects in others. Examples are the polymorphisms of cytochrome P450. Children who were treated for relapse of acute lymphoblastic leukemia who had the wild-type variant of CYP1A1 had a better outcome than those with the variant type. Another important example for differential actions of drugs is thiopurine methyltransferase deficiency, where children with this defect tolerate only minimal doses of thiopurines.

These examples illustrate that detailed and comprehensive genetic information on drugs metabolism should be available in order to find the optimal individual dose for a patient.

Likewise genetic information on cancer cells and/or host interactions should be known and used to determine the optimally effective drugs or the individually required drug dose. Examples of new drug developments or potentials are tyrosine kinase inhibitors (Imatinib, Gleevec). More recently a much more potent inhibitor active even against multiple kinases as well as receptor tyrosine kinases (dasatinib, BMS-354825) has been developed and is being studied in clinical trials (Carter et al. 2005; Hochhaus 2005). Histone deacetylase inhibitors (HDACi) are agents that have completely different modes of action than commonly used cytostatic drugs. Angiogenesis inhibitors may be capable of preventing metastases of tumors and thus change the disease to a localized instead of a generalized, systemic process. Small RNA (RNAi) may be used to alter the genome of tumor cells and stop proliferation.

The development and availability of such compounds would lead to a completely different understanding of cancer and might even change the paradigm of radical eradication of cancer cells to growth control.

In summary, research in children is essential and urgently required. Research in children means research for children, which should best be done in the form of controlled clinical trials. This will offer the chance to obtain information on pharmacokinetics and pharmacogenetics and to make optimal use of pharmacological data. In the future, this information will be translated into individualized and thus more effective treatments with fewer acute side effects and fewer adverse late sequelae, better survival, and improved quality of life.

References

Anonymous (2006) Deutsches Kinderkrebsregister: Jahresbericht 2004 (1980–2003). http://www.kinderkrebsregister.de/:, Cited 31 Mar 2006

Boos J, Werber G, Ahlke E, Schulze-Westhoff P, Nowak-Gottl U, Wurthwein G, Verspohl EJ, Ritter J, Jurgens H (1996) Monitoring of asparaginase activity and asparagine levels in children on different asparaginase preparations. Eur J Cancer 32A:1544–1550

Bostrom BC, Sensel MR, Sather HN, Gaynon PS, La MK, Johnston K, Erdmann GR, Gold S, Heerema NA, Hutchinson RJ, Provisor AJ, Trigg ME; Children's Cancer Group (2003) Dexamethasone versus prednisone and daily oral versus weekly intravenous mercaptopurine for patients with standard-risk acute lymphoblastic leukemia: a report from the Children's Cancer Group. Blood 101:3809–3817

Brochhausen C, Seyberth HW (2000) Pharmakotherapie in der Pädiatrie – Besonderer Schutz, besondere Herausforderung. Innovartis 2000 www.studentenservice.novartis.de/content_pdf/innovartis/i0013135.pdf. Cited 31 Mar 2006

Carter TA, Wodicka LM, Shah NP, Velasco AM, Fabian MA, Treiber DK, Milanov ZV, Atteridge CE, Biggs WH 3rd, Edeen PT, Floyd M, Ford JM, Grotzfeld RM, Herrgard S, Insko DE, Mehta SA, Patel HK, Pao W, Sawyers CL, Varmus H, Zarrinkar PP, Lockhart DJ (2005) Inhibition of drug-resistant mutants of ABL, KIT, and EGF receptor kinases. Proc Natl Acad Sci U S A 102:11011–11016

Conroy S, McIntyre J, Choonara I (1999) Unlicensed and off label drug use in neonates. Arch Dis Child Fetal Neonatal Ed 80:F142–F145

Conroy S, Choonara I, Impicciatore P et al (2000) Survey of unlicensed and off label drug use in paediatric wards in European countries. BMJ 320:79–82

Conroy S, Newman C, Gudka S (2003) Unlicensed and off label drug use in acute lymphoblastic leukaemia and other malignancies in children. Ann Oncol 14:42–47

Creutzig U, Henze G, Bielack S, Herold R, Kaatsch P, Klussmann JH, Graf N, Reinhardt D, Schrappe M, Zimmermann M, Jürgens H (2003) Krebserkrankungen bei Kindern: Erfolg durch einheitliche Therapiekonzepte seit 25 Jahren. Deutsches Ärzteblatt 100:A842

Creutzig U, Zimmermann M, Hannemann J, Kramer I, Pfistner B, Herold R, Henze G (2005) Quality management for clinical trials within the German Competence Network Paediatric Oncology and Haematology. Onkologie 28:333–336

Duval M, Suciu S, Ferster A, Rialland X, Nelken B, Lutz P, Benoit Y, Robert A, Manel AM, Vilmer E, Otten J, Philippe N (2002) Comparison of Escherichia coli-asparaginase with Erwinia-asparaginase in the treatment of childhood lymphoid malignancies: results of a randomized European Organisation for Research and Treatment of Cancer – Children's Leukemia Group phase 3 trial. Blood 99:2734–2739

Hochhaus A (2005) Phase III Studie zum Nachweis der Aktivität von BMS-354825 bei Patienten mit Ph-positiver CML in chronischer Phase nach Resistenz auf Imatinib (400 mg/Tag). http://www.kompetenznetz-leukaemie.de/kn_home/Studien/studie_136.html. Cited 31 Mar 2006

McIntyre J, Conroy S, Avery A et al (2000) Unlicensed and off label prescribing of drugs in general practice. Arch Dis Child 83:498–501

Merck & Co., Inc. (1999) Drug treatment in newborns, infants, and children: drug doses. In: Beers MH, Berkow R (eds) The Merck manual of diagnosis and therapy. Merck Research Laboratories, Whitehouse Station, NJ, pp 2117–2118. www.merck.com/mrkshared/mmanual/section19/chapter258/258b.jsp. Cited 31 Mar 2006

Turner S, Gill A, Nunn T et al (1996) Use of "off label" and unlicensed drugs in paediatric intensive care unit. Lancet 347:549–550

Turner S, Longworth A, Nunn AJ, Choonara I (1998) Unlicensed and off label drug use in paediatric wards: prospective study. BMJ 316:343–345

Vieira Pinheiro JP, Ahlke E, Nowak-Gottl U, Hempel G, Muller HJ, Lumkemann K, Schrappe M, Rath B, Fleischhack G, Mann G, Boos J (1999) Pharmacokinetic dose adjustment of Erwinia asparaginase in protocol II of the paediatric ALL/NHL-BFM treatment protocols. Br J Haematol 104:313–320

9 Integration of Pediatric Aspects into the General Drug Development Process

K. Rose

Abstract. Drug treatment of children is today less regularly based on formal clinical testing than adults. This has led to concerns regarding the safety and efficacy of pediatric medicines and resulted in public action in the United States and the European Union. The reasons for the increasing awareness include better understanding of child physiology, increased trust in GCP (good clinical practice), improved treatment of several severe childhood diseases, a changed view of the child as a subject in society, and more. The US has successfully introduced pediatric legislation that facilitates participation of children in pharmaceutical innovation, and comparable approaches are now being discussed in Europe and Japan. While the outcome of the EU pediatric regulation in the near future is still open, the US pediatric legislation has been highly successful over the past 8 years and will be revised before it expires in September 2007. Innovative drugs are today being developed by global pharmaceutical companies. Adding pediatric aspects to this development process is a complex task where companies need to build up internal competency. Bureaucratic procedures that could be harmful to the companies' economic fundaments need to be avoided, and an appropriate ethical framework is required. This needs to be addressed by all partners in healthcare, including regulatory authorities, the pharmaceutical industry, pediatricians, patients and others in a sense of shared responsibility.

9.1 Clinical Drug Development is a Young Discipline

Clinical testing of new medications is historically quite new. At the beginning of the twentieth century it was legal in the United States to claim therapeutic efficacy for any concoction, including pseudo-medicines that promised to cure cancer, arthritis, or tuberculosis. If the product was simply ineffective, it "only" prevented the consumer from purchasing effective treatment. However, often these so-called wonderdrugs were dangerous and could have severe side effects, including permanent disability or death (Hilts 2003). It took several major tragedies, specifically the deaths following the sulfanilamide elixir disaster in 1937 (Wax 1995) or the thousands of children with phocomelia in 1960–1962 following intake of thalidomide by their mothers (Taussig 1962) to mobilize the public opinion sufficiently to introduce requirements, step by step, that drugs needed proof of safety and efficacy (Hilts 2003). Good clinical practice (World Medical Association 2004; International Conference of Harmonisation 1996) has evolved over the ensuing decades as a framework that regulates the performance of clinical testing in humans.

9.2 Off-Patent Use of Medicines in Children

When proof of safety and efficacy by clinical testing became mandatory, clinical trials were conducted in adults, usually in healthy male volunteers. To avoid being sued, drug producers explicitly stated that the product had not been tested in children and therefore could not take any responsibility when the respective product was used in children. This pediatric disclaimer was observed by Shirkey in 1963, and he defined children as "therapeutic orphans" (American Academy of Pediatrics 1996; Shirkey 1999). Frequently, medicines are given to children off-label. This transfers the responsibility and liability from the drug producer to the treating pediatrician or medical doctor, who is faced with the dilemma of either not treating the child with a medication or treating the child but carrying the risk of prescribing doses that are too high or too low. Pediatricians started rather early to lobby for more and better drug testing in the pediatric population, but little changed in the three following decades. In most countries, one or several formulas were developed to assess doses necessary in children, usually extrapolating the dose for a child on the basis of the child's weight, body surface, or on the combination of both (Royal College of Paediatrics and Child Health 2003; Von Harnack 2003). Numerous publications describe the extent of off-label use in several European countries (Choonara et al. 2000; McIntyre et al. 2000; Turner et al. 1998; Schaad 2001). Today, off-label use in children ranges between 10% of drugs prescribed in general practice to up to 90% of drugs prescribed in neonate intensive wards.

9.3 US Pediatric Legislation and the Emerging EU Pediatric Regulation

The first successful attempt to facilitate the generation of pediatric clinical data was in the US when the FDA Modernization Act (FDAMA) was signed in 1997, which introduced a voluntary incentive for the pharmaceutical industry (Food and Drug Administration 1997). The incentive was offered in exchange for a pediatric development plan that had to be agreed upon with the FDA. This plan is called the "Written Request." In the early years after 1997, the FDA issued many written requests in general terms. A company that committed in writing to the request could expect a six-month extension of market exclusivity to its entire product. "Market exclusivity" means that the pediatric exclusivity is added to whichever protection a specific drug has against generic competition. Mostly this is the patent itself, but there are several other forms of market exclusivity. The incentive interests companies because generic competition is very strong in the US. The loss of sales from the moment of the loss of the patent on can be up to 95%. Therefore, protection against generic erosion can be a considerable incentive. The concept of pediatric exclusivity proved so successful that in 2002 both houses of Congress in the US unanimously prolonged it under the name "Best Pharmaceuticals for Children Act" (BPCA) until the end of September 2007 (US Congress 2001).

The clinical investigations initiated by FDAMA and BPCA were initially pharmacokinetics/pharmacodynamic (PK/PD) data, extrapolation on dosing in smaller age groups, and new indications outside the mainstream use of the drug in the adult population. Two examples are the investigation of bisphosphonates (alendronate, risendronate, zoledronate) in the treatment of osteogenesis imperfecta, or the use of investigation in the therapeutic potential of tamoxifen in the treatment of McCune-Albright-Syndrome (Food and Drug Administration 2005). In the last few years, the FDA has also increasingly asked for the development of suitable pediatric formulations.

The other side of the US pediatric legislation was initially called the "pediatric rule" and should give the FDA authority to request pediatric studies from pharmaceutical companies (Food and Drug Administration 1998). This pediatric rule was struck down by a federal court in 2000. In December 2003, it was reintroduced as a law under the name Pediatric Research Equity Act (PREA, US Congress 2003). It gives the FDA authority to request pediatric assessments of a new drug in early development stages as well as request mandatory clinical trials in children in the same indication as in adults. The FDA has started to issue requests for pediatric studies in those drugs where companies have rejected the originally issued written request.

Both PREA and BPCA are explicitly linked, i.e., the FDA and pharmaceutical companies are encouraged to negotiate a compromise where pediatric development is agreed upon and will then be rewarded with additional an 6 months market exclusivity at the end of patent life. As both laws will expire September 30, 2007, this gives pharmaceutical companies the opportunity to negotiate a written request until the end of September 2007 and then execute the agreed upon pediatric studies within the agreed upon time frame, which can be several years later than 2007.

9.4 Timing of Pediatric Development, Deferrals, and Waivers

In article 4 §1 of the EU consultation paper (EU Commission March 2004) the submission of the *results* of all studies performed and details of all information collected in accordance with a completed, agreed pediatric investigation plan are listed as an explicit condition for applying for a marketing authorization of a new medicine, unless a deferral or a waiver has been granted. The wording in the draft regulation proposed September 2004 is comparable (EU Commission September 2004). In the debate following these proposals, it was emphasized by the pharmaceutical industry that results of pediatric studies will practically never be available when adult data are submitted for a marketing authorization (European Federation of Pharmaceutical Industries and Associations 2005) The principal guideline is here ICH E 11 "Principles for Clinical Evaluation of Medicinal Products in the Pediatric Population" (International Conference of Harmonisation 2000). Its main scenarios are (1) serious and life-threatening diseases that affect primarily children, (2) serious and life-threatening diseases that affect both adults and children, and (3) all other diseases. Scenarios 1 and 2 would justify early development in children, while for all other diseases and conditions pediatric data are usually generated at a later time point. Even for serious and life-threatening diseases that exist both in adults and children, i.e., scenario 2, ICH E 11 asks for pediatric development to begin early in the pediatric population *after following assessment of initial safety data and reasonable evidence of potential benefit* (International Conference of Harmonisation 2000, 2.3.2).

Exposure of children to a new, mainly untested substance can ethically be justified only where the disease poses danger of life or of severe damage to health without a therapeutic alternative available. For most diseases, there are already suitable therapeutic options available. Scenario 2 will most probably be only invoked in the case of therapeutic breakthroughs in diseases that until now were incurable. Therefore, availability of the *results* of clinical studies in children at the time point of registration for the adult population may be the wishful thinking of politicians, but would in reality endanger a timely registration of new medicines in adults.

A decision to go for scenario 2, i.e., initiate early clinical development in children, has far-reaching consequences within the developing company. Several development blocks have to be shifted to earlier stages, mainly preclinical safety and toxicology and the development of a pediatric formulation. Many projects are abandoned at these early development stages. An increased generation of preclinical safety and tolerability data and of pediatric formulations for drugs that are later abandoned and not further developed would only increase the financial burden of drug development in general and would benefit not a single child. A decision to opt for early pediatric development requires careful balance of the potential therapeutic value, ethical concerns, and inherent risks.

It is likely that in most cases the authorities will issue a waiver at the time of submission for a market authorization for adults for those age groups in children that do not have the targeted disease, and will issue a deferral where the disease in children exists. The time frame of deferrals issued by the FDA has until now been 5 years. Usually the pediatric clinical trials start in the second half of this time frame.

9.5 Integrating Pediatric Aspects into the General Drug Development Process

Modern drug development has evolved into a complex global, interdisciplinary, complex, and highly regulated process. It reaches from basic research where new chemical or biological entities are discovered or designed and comprises the known development phases I–III until submission of the dossier to the health authorities, and continues after registration in the first country as phase IV studies and postmarketing surveillance. Pediatric drug development encompasses all these development phases, and quickly reviewing each of the phases shows how much further we will have to go.

At the level of basic research there is little child-specific knowledge. We do not have animal models for diseases in children. We know very little about which receptors and pathways are used and switched on and off in the child's body. However, we are starting to broaden at least our vision toward what will be possible in the future. ICH E 11 only gives two examples for its scenario 1, i.e., life-threatening and serious diseases ("medicinal products for diseases predominantly or exclusively affecting pediatric patients"; International Conference of Harmonisation 2000, 2.3.1): "surfactant for respiratory distress syndrome in preterm infants and therapies targeted at metabolic or genetic diseases unique to the pediatric population." But today we can go one step further. Many autoimmune diseases start in childhood, such as asthma, allergic rhinitis, atopic dermatitis, or insulin-dependent diabetes (Kulmala 2003). These diseases are indeed children's diseases, but as the affected patients mostly survive, we know them well from the continuous treatment in the adult population. Only today have we started to think about the possibility of interrupting the disease cascade in childhood. Eventually, there will be clinical trials to modify the course of disease or even to prevent them completely, and these trials will have to be conducted in children.

At the point in time where in development suitable targets are selected for a new compound, a great deal of additional knowledge will now be required on the epidemiology of diseases in childhood and to what degree diseases we know in adults are the same ones, comparable, or completely different when manifested in children.

Clinical pharmacology plays a key role in pediatric drug development. Key questions are the classical questions of ADME (absorption, distribution, metabolization, excretion), the relationship between serum concentration and therapeutic efficacy, models and simulations that help to extrapolate from adults to adolescents and then cascading down from older children to the younger and very young age groups. Surrogate markers play an increasing role in general drug development and will have a special place in pediatric development.

The challenges of pediatric drug development in the late clinical phases are multiple again. Investigators need experience with pediatric clinical trials, they need rooms that are sufficiently equipped not only for the child, but for parents and siblings who might accompany the child. Experienced research personnel are essential. Stopping rules need to be established when blood withdrawal fails. The laboratory involved must be experienced in handling small blood volumes.

9.6 Building up Pediatric Competency in Pharmaceutical Companies

Companies will need to build up the necessary expertise in-house or by relying on external specialists. Pediatric contract research organizations (CROs) have started to offer their services. But even if all necessary expertise can be bought externally, minimal competence is required in each company for establishing company-specific guidelines, to give the necessary input to all the departments that are exposed to pediatric questions, to supervise CROs, to interpret data from clinical development, and to understand and discuss pediatric issues with the regulatory authorities.

9.7 Will the EU Draft Regulation Leading to More Pediatric Research in the Near Future?

The debate on a European contribution to childhood health has until now focused very much on the financial incentive as it is granted by the US authorities. EU health authorities requested quite early access to the data generated by pediatric research projects, which were driven by the US pediatric exclusivity reward. Not all pharmaceutical companies complied immediately. This was criticized by the EU Commission in the introduction to their consultation paper in March 2004 (EU Commission, March 2004). Later, EFPIA (European Federation of Pharmaceutical Industries and Associations, representing the European research-based pharmaceutical industry) asked its members to share existing pediatric data with the health authorities.

While the debate so far has mainly focused on access to results from pediatric research done in the past, to what extent the EU regulation will stimulate pediatric research on its own remains unknown. Once the regulation will be in force, possibly around 2007, the newly implemented EU Pediatric Committee (PC) will be able to give input. Many new drugs target a disease for which there is already an established medical cure, but they offer additional benefit compared to the existing therapeutic options. For such conditions, it would not be ethical to expose children prematurely to substances for which only very limited experience is available in adults. True therapeutic breakthroughs that save lives of patients with so far almost incurable diseases remain the exception. Therefore, it can be expected that for most new medicines a waiver or a deferral will be issued allowing for collection of more safety and efficacy data in adults before children are exposed to the new compound. The usual timeframe for deferrals in the US is at present 5 years: after 5 years, results from clinical studies in children are expected. In consequence, clinical studies in children will start 2–3 years after adult registration. Following this path, the very first pediatric trials triggered by the EU regulation will not start before 2007 + 3, i.e., 2010.

For the EU regulation to stimulate pediatric research in Europe in the near future, it needs to encourage pediatric research on the modern medicines used today. For most of these drugs, the US legislation has already resulted in a first pediatric dataset. Such data include data on dosing in lower age groups, PK/PD data, sometimes use in new indications, and sometimes a pediatric formulation (Food and Drug Administration 2005). The areas where additional pediatric research could be stimulated by the EU regulation would be observations over longer time, investigations into more subpopulations, additional new indications, and many more. For Europe to start playing a major role again in global pediatric research, it will be important to admit that for the time being the most immediate unmet medical needs have to a large degree already been covered by the US pediatric legislation.

This debate has started in Europe, but until now only on a quite limited level. It has begun to reach the national health authorities and those associates in research and development that have a very special interest in children. Hopefully this debate will penetrate deeper over the coming years.

9.8 Globalization of Clinical Research and Europe's Competitiveness

Even though the political level the EU expresses the intention to play a major role in pediatric research, Europe's competitiveness in general in comparison to the US and other regions has declined over the last decade. While our continent struggles to find a suitable way to meet the Lisbon agenda, other regions are preparing themselves for future competition. For example, in India a counterpart to EFGCP (EFGCP 2005) is being built up. Most Indian doctors are fluent in English. In other regions, people are less used to relying completely on their governments. This may become an advantage for those research centers and structures that prove to be sufficiently business-oriented to offer a good partnership with industry. Increased demand for pediatric research in the US and EU will not automatically lead to research performed in the US and EU. As long as the quality of any data regardless of their region of origin is guaranteed through strict adherence to GCP and as long as the FDA and EMEA can audit the investigational sites and the source data, such data will be accepted by the EMEA and FDA. A European pediatric research network will have to compete from the beginning with other regions of the world. With increasing capabilities of information technology, this trend will become even stronger in the near future

9.9 Conclusions

We should take the evolving EU pediatric regulation as a paradigm for Europe's competitiveness in the modern world. The first steps have been made by the US. There is much more work to be stimulated by Europe and in Europe. This continent has many resources, educated people, culture, and all the other ingredients needed. In the US, the general climate is more favorable to business, and here we can still learn a lot as Europeans. Improving children's health and the necessary research to ensure this should be addressed in a spirit of partnership of all involved parties.

References

American Academy of Pediatrics (1996) American Academy of Pediatrics, Committee on Drugs: Unapproved uses of approved drugs: the physician, the package insert, and the Food and Drug Administration: subject review. Pediatrics 98:143–145

Belay ED, Bresee JS, Holman RC et al (1999) Reye's syndrome in the United States from 1981 through 1997. N Engl J Med 340:1377–1382

Caldwell PHY et al (2004) Clinical trials in children. Lancet 364:803–811

Choonara I et al (2000) Paediatric medicines: global development and clinical investigations. Scrip Report. PJB Publications, London

Conroy S, McIntyre J, Choonara I (1999) Unlicensed and off label drug use in neonates. Arch Dis Child Fetal Neonatal Ed 80:F142–F145

Donat JF, Bocchini JA Jr, Gonzales E et al (1979) Valproic acid and fatal hepatitis. Neurology 2:273–274

EFGCP (2005) www.efgcp.org

European Federation of Pharmaceutical Industries and Associations (2004) EFPIA welcomes the commission proposal to encourage research and development of medicines for the benefit of children. http://www.efpia.org/3_press/20040929.htm. Cited 6 April 2006

European Federation of Pharmaceutical Industries and Associations (2005) www.efpia.org. Cited 6 April 2006

EU Commission (March 2004) Commission consultation on a draft proposal for a European Parliament and Council Regulation (EC) on medicinal products for paediatric use. http://pharmacos.eudra.org/F2/pharmacos/docs/Doc2004/mar/Paediatric%20consultation%20document%20final%208%20March%2004.pdf. Cited 14 July 2005

EU Commission (September 2004) Proposal for a regulation of the European Parliament and of the council on medicinal products for paediatric use. http://pharmacos.eudra.org/F2/Paediatrics/docs/Paeds%20reg%20adopted%2029%20September%202004%20English.pdf. Cited 14 July 2005

Food and Drug Administration (1997) The pediatric exclusivity provision January 2001 – Status Report to Congress. http://www.fda.gov/cder/pediatric/reportcong01.pdf. Cited 14 July 2005

Food and Drug Administration (1998) Regulations requiring manufacturers to assess the safety and effectiveness of new drugs and biological products in pediatric patients: final rule. http://www.fda.gov/ohrms/dockets/98fr/120298c.txt. Cited 14 July 2005

Food and Drug Administration (2005) BPCA / Pediatric Exclusivity Statistics.
http://www.fda.gov/cder/pediatric/labelchange.htm;
http://www.fda.gov/cder/pediatric/breakdown.htm;
http://www.fda.gov/cder/pediatric/wrlist.htm;
http://www.fda.gov/cder/pediatric/wrstats.htm;
http://www.fda.gov/cder/pediatric/wrstats.htm;
http://www.fda.gov/cder/pediatric/specdis.htm;
http://www.fda.gov/cder/pediatric/Summaryreview.htm. Cited 14 July 2005

International Conference of Harmonisation (1996) ICH E 6 Guideline for good clinical practice. http://www.ich.org/MediaServer.jser?@_ID=482&@_MODE=GLB. Cited 14 July 2005

International Conference of Harmonisation (2000) ICH E 11 Clinical Investigation of Medicinal Products in the Pediatric Population. http://www.ich.org/MediaServer.jser?@_ID=487&@_MODE=GLB. Cited 14 July 2005

Impiccatore P, Choonara I (1999) Status of new medicines approved by the European Medicine Evaluation Agency regarding paediatric use. Br J Clin Pharmacol 49:93–97

Kulmala P (2003) Prediabetes in children. Natural history, diagnosis, and preventive strategies. Pediatr Drugs 5:211–221

McIntyre J et al (2000) Unlicensed and off label prescribing of drugs in general practice. Arch Dis Child 83:498–501

Parke TJ, Stevens JE, Rice AS et al (1992) Metabolic acidosis and fatal myocardial failure after propofol infusion in children: five case reports. BMJ 305:613–616

Rose K (2005) Pediatric drug development. Implementation of pediatric aspects into the general drug development process. App Clin Trials 14:50–53 (http://www.actmagazine.com/appliedclinicaltrials/article/articleDetail.jsp?id=140819. Cited 14 July 2005)

Royal College of Paediatrics and Child Health (2000) Royal College of Paediatrics and Child Health, Ethics Advisory Committee Guidelines for the ethical conduct of medical research involving children. Arch Dis Child 82:177–182

Royal College of Paediatrics and Child Health (2003) Medicines for children. http://www.rcpch.ac.uk/publications/formulary_medicines.html. Cited 14 July 2005

Schaad UB (2001) Drug therapy in children: still more art than science. Curr Opin Infect Dis 14:301–302

Shirkey H (1999) Editorial comment: therapeutic orphans. Pediatrics 104 [Suppl]:pp 583–584

Smith K (1998) Registering paediatric medicines in Australia. Regulatory Affairs J 9:304–308.

Taussig HB (1962) A study of the German outbreak of phocomelia. JAMA 180:1106–1114

Turner S et al (1998) Unlicensed and off label drug use in paediatric wards: prospective study. BMJ 316:343–345

Uchiyama A (2002) Pediatric clinical studies in Japan: regulations and current status. App Clin Trials 11:pp 57–59 (http://www.actmagazine.com/appliedclinicaltrials/article/articleDetail.jsp?id=83734. Cited 14 July 2005)

US Congress (2001) Pediatric rule. http://www.fda.gov/cder/pediatric/PL107–109.pdf. Cited 14 July 2005

US Congress (2003) Best pharmaceuticals for children act. http://www.fda.gov/cder/pediatric/S-650-PREA.pdf. Cited 14 July 2005

Von Harnack GA (2003) Pädiatrische Dosistabellen, ISBN 3804715877

Wax P (1995) Elixirs, diluents, and the passage of the 1938 Federal Food, Drug and Cosmetic Act. Ann Intern Med 122:456–461

World Medical Association (2004) World Medical Association Declaration of Helsinki. Ethical principles for medical research involving human subjects. http://www.wma.net/e/policy/pdf/17c.pdf. Cited 14 July 2005

Wilson JT (1999) An update on the therapeutic orphan. Pediatrics 104:585–590

10 Current Stumbling Blocks in Oncology Drug Development

C.D. Gimmi

Abstract. The prognosis of patients with metastatic cancer remains poor and treatment strategies including newer generations of chemotherapeutics have not significantly improved survival in most solid tumors. New approaches are required to further improve patient outcome and survival. Recently, a major leap in the understanding of the molecular mechanisms involved in signal transduction pathways that contribute to tumor growth have been identified as therapeutic targets. Particularly molecules involved in cellular proliferation (e.g., tyrosine kinases) and angiogenesis have been considered as targets for new treatment approaches. Novel therapeutics that specifically target kinase transduction pathways have shown promise as single agents and in combination with standard chemotherapy. In addition, results of recent studies with antiangiogenic monoclonal antibodies validate the use of this class of targeted therapeutics as an important new treatment modality in cancer. This review will focus on the drug development stumbling blocks of targeted treatment modalities in cancer.

10.1 Introduction

The traditional cytotoxic agents, most of which act on DNA or tubulin, are active in a multitude of different tumor types. Preclinical studies have demonstrated that there is a direct relationship between chemotherapy dose and tumor cell kill; however, these preclinical models did not predict the clinical outcome. The traditional design of early clinical trials of cytotoxic agents was based on the selection of clinical endpoints for benefit prediction. Although phase III trials involving cytotoxic compounds used predetermined safety and efficacy rules in planned early review, the clinical benefit remained unpredictable.

The emergence of targeted agents has added new challenges to oncology drug development. Unlike cytotoxic agents, these compounds have myriad targets, including growth factor receptors, components of signaling transduction pathways, cell cycle regulator proteins, gene transcription factors, and proteins regulating angiogenesis. The deployment of traditional dosage, dose regimen findings, and endpoints to targeted treatments may have to incorporate new surrogates for dose selection or activity evaluation, as neither toxicity nor tumor shrinkage may be expected with these agents. An additional challenge in the development of targeted therapy may become the integration of new drugs into the standard of treatment and the acceptance by the regulatory environment.

10.2 Multitude of Targets: How to Select a Target?

During the last two decades, cancer biology has gained understanding and identified new molecular pathways and genetic alterations that cause malignant transformation, invasion, angiogenesis, and metastases. Several coordinated molecular events have been identified that lead to cancer development and have become attractive targets for the development of anti-cancer treatment:

- Altered gene activity (including oncogene activation or loss of tumor suppressor gene activity, e.g., *p*53) reduces cell cycle control and susceptibility to apoptosis

- Constitutive activation of growth factor (GF) receptors independent of GF
- Deregulated secretion or responsiveness to exogenous (paracrine) or autocrine GF, leading to autonomous cell growth and proliferation
- Novel or enhanced secretion of angiogenic growth factors
- Enhanced production and secretion of matrix metalloproteinases (MMP) and other enzymes that interact with the extracellular matrix (ECM), to promote local invasion into tissues and facilitate metastases
- Repair of DNA damage may not occur, leading to more genetic alterations that drive malignant transformation

The epidermal growth factor receptor (EGFR) is an ideal target for cancer therapy because it is commonly expressed at a high level in a variety of solid tumors and it has been implicated in the control of cell survival, proliferation, metastasis and angiogenesis. The extracellular, ligand-binding domain has the capability to bind a number of regulatory factors such as EGF, TGF-alpha and to promote EGFR dimerization and autophosphorylation. The cytoplasmic domain containing the tyrosine kinase catalytic activity responds to the receptor autophosphorylation by increasing tyrosine kinase activity and phosphorylation of other tyrosine motifs. These tyrosine motifs are the binding sites for a cascade of intracellular signaling pathways such as PLC-gamma, Ras-Raf-MEK-MAP kinase, PI3K, Akt, etc. and result in functions such as cellular proliferation, malignant transformation, migration, and anti-apoptosis (Ciardello and Tortora 2003) (Fig. 1).

Agents targeting molecular mechanisms are expected to be more cytostatic than cytotoxic and to be more tumor-specific and therefore spare significant organ toxicity. These agents may become complementary to established cytotoxic drugs and improve efficacy further beyond standard treatment due to nonoverlapping toxicity profiles.

10.3 Validation of a Target

A few targets have been adequately validated in oncology. However, the presence of redundant and compensatory cellular pathways, innate and acquired drug resistance, and genetic instability of tumors can decrease the clinical value of a particular target. There is a growing consensus to

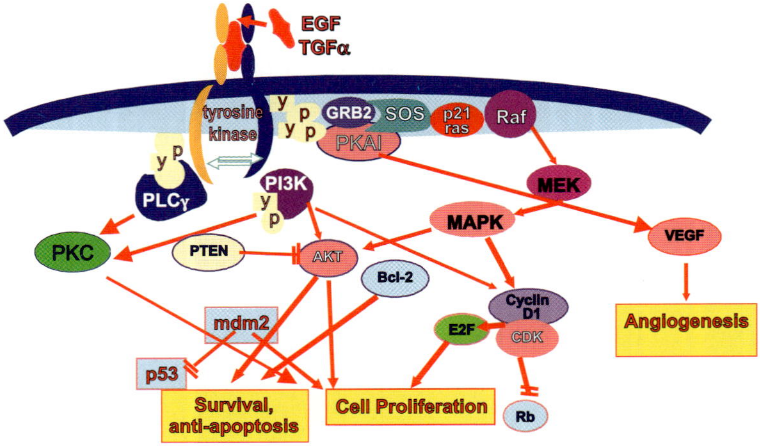

Fig. 1. Simplified overview of the molecular pathways leading to tumor formation. Growth factor signals are propagated from the cell surface, through the action of transmembrane receptors, to intracellular effectors that control critical functions in human cancer cells, such as survival, differentiation, cell proliferation, angiogenesis, and inhibition of apoptosis

consider a target valid if (1) it is both uniquely and markedly overexpressed in a specific type of cancer (tumor-specific); or (2) it is essential for the survival of the cancer in question (rate-limiting) (Roberts et al. 2003).

1. The validity of human epidermal growth factor-2 (HER-2) as a target has been established successfully in trials showing objective responses with an anti-HER2 monoclonal antibody (trastuzumab, Herceptin) as a single agent in patients with metastatic breast cancer.

In a first study, trastuzumab was evaluated in 222 women with advanced metastatic disease who had received extensive prior therapy (Cobleigh et al. 1999). Only patients who showed moderate (2+) or strong (3+) immunohistochemical staining for HER-2 in more than 10% of tumor cells were eligible. Trastuzumab administered as a single agent produced eight complete and 26 partial responses, for an objective response rate of 15% in the intent-to-treat population (95% confidence interval

[CI], 11%–21%). The median duration of response was 9.1 months; the median duration of survival was 12.8 months. Similar results were seen in 111 women with HER-2-overexpressing metastatic breast cancer who were randomized to receive first-line treatment with trastuzumab (Vogel et al. 2002). The objective response rate was 30% (95% CI, 18.2%–34.4%), with seven complete and 23 partial responses. Response rates of the 111 assessable patients with 3+ and 2+ HER-2 overexpression by immunohistochemistry (IHC) were 35% (30/84) and none (0/27), respectively. The clinical benefit (objective responses and stable disease) rates in assessable patients with 3+ and 2+ HER-2 overexpression were 48% and 7%, respectively. These data suggested that patients with a score 3+ had a greater benefit from the treatment with trastuzumab than patients with a score 2+. Further information suggested that there was a high degree of concordance between gene amplification and a score 3+ in immunohistochemistry for HER-2. In this same study, 108 assessable patients were retrospectively analyzed for HER-2 gene amplification by fluorescence in situ hybridization (FISH) (Vogel et al. 2002). The objective response rate was 34% in patients with FISH + tumors vs 7% for patients with FISH – tumors. Taking these facts together, the efficacy of trastuzumab depends on the degree of HER-2 expression and the gene amplification by FISH in tumors. This improvement of clinical benefit was maintained in breast cancer patients with FISH + tumors (amplified HER-2 gene) when trastuzumab was combined with chemotherapy (Slamon et al. 2001; Baselga 2001; Mass et al. 2000). A significant improvement of progression-free survival (PFS, HR $= 0.7$) was seen in this patient population when treated with chemotherapy and trastuzumab vs chemotherapy alone. In contrast, no difference in progression-free survival was seen in the patient population with FISH-negative tumors between the two treatment arms.

However, the validation of a target requires more than the use of a drug in a preselected patient population, it requires clinical validation of the targeted therapy. There are multiple ways of intercepting with a growth factor receptor, and each of these mechanisms has the potential to elicit a different clinical efficacy. For instance, trastuzumab binds to the juxtamembrane epitope of HER-2 extracellular domain in a region that is normally the site of ectodomain cleavage. This cleavage constitutes one mode of HER-2 kinase activation that is blocked by trastuzumab.

In contrast, another HER-2-targeting, humanized monoclonal antibody, pertuzumab (Omnitarg), binds to a distinct epitope that sterically hinders the HER-2 dimerization motif, resulting in blockade of HER-2 dimerization with other HER receptors. This activity is not dependent on HER-2 overexpression, as was demonstrated in a large body of preclinical experiments (Nahta et al. 2004). So far the preclinical efficacy could not be translated into clinical benefit, as demonstrated in randomized phase II trail where patients with metastatic breast cancer with low HER-2 expression were randomized to either initial loading dose of pertuzumab 840 mg followed by 420 mg every 3 weeks (arm A), or 1,050 mg every 3 weeks (arm B) (Cortes et al. 2005). In total, two of 78 patients achieved a PR for an ORR of 3% and six of 78 patients (8%) had SD for 6 months. The non-crossreactivity of these two antibodies suggests that the combination of the two anti-HER-2 antibodies may be more effective than treatment with a single anti-HER-2 antibody (Nahta et al. 2004).

2. The importance of developing agents against targets that are not only uniquely expressed in tumors, but are also essential for the survival of the cancer in question and therefore for the clinical outcome has become increasingly clear.

Vascular endothelial growth factor (VEGF) is a key regulator for angiogenesis and is essential for tumor growth and metastasis. The biological effects of VEGF are mediated by two receptor tyrosine kinases (RTKs), VEGFR-1 and VEGFR-2, which differ considerably in signaling properties. Vascular endothelial growth factor is one of the best characterized pro-angiogenic growth factors, and a specific mitogen for endothelial cells, with a key role in activating the angiogenic switch. VEGF is responsible for endothelial cell survival signaling in newly formed vessels and induces significant vascular permeability. VEGF appears to be a critical factor in tumor angiogenesis. Currently, several VEGF inhibitors are undergoing clinical testing in several malignancies (Ferrara et al. 2003). Bevacizumab (Avastin), a humanized monoclonal antibody developed against VEGF, has been shown to prevent the interaction of VEGF with its receptors, VEGF receptor-1 and VEGF receptor-2, on the surface of vascular endothelial cells. This interaction with the VEGF receptors inhibits the downstream signaling pathways that would normally lead to growth, proliferation, migration, and survival of endothelial cells.

Bevacizumab was investigated in a randomized phase III trial of patients with previously untreated advanced colorectal cancer. A total of 813 patients were randomized to bevacizumab (5 mg/kg, IV, biweekly) in combination with irinotecan, bolus fluorouracil, and leucovorin (IFL), compared to IFL alone. The median duration of survival was 20.3 months in the group given IFL plus bevacizumab, as compared with 15.6 months in the group given IFL plus placebo, corresponding to a hazard ratio for death of 0.66 ($p < 0.001$). The median duration of progression-free survival was 10.6 months in the group given IFL plus bevacizumab, as compared with 6.2 months in the group given IFL plus placebo (hazard ratio for disease progression, 0.54; $p < 0.001$). This study demonstrated that the addition of bevacizumab to fluorouracil-based combination chemotherapy results in statistically significant and clinically meaningful improvement in survival among patients with metastatic colorectal cancer (Hurwitz et al. 2004).

In another phase III trial, patients with metastatic breast cancer after anthracycline and taxane failures were randomized to chemotherapy with capecitabine ($n = 431$) alone vs capecitabine with bevacizumab ($n = 431$). In this phase III trial, the median PFS was 4.17 months for capecitabine monotherapy vs 4.86 months for capecitabine + bevacizumab (HR = 0.98), although the overall response rate was significantly higher (19.8%) with the combination vs 9.1% for monotherapy (Miller et al. 2005). One explanation for this controversy may be the distribution of pro- vs antiangiogenic factors, e.g., bFGF, TGFβ-1, hypoxia-inducible factors, and platelet-derived endothelial cell growth factor (PD-ECGF; also known as thymidine phosphorylase), which may become essential in later stages of tumor progression, whereas small tumors may recruit blood vessels mainly through secreting VEGF. This suggests that while early-stage tumors may express VEGF as a principle pro-angiogenic factor, angiogenesis in late-stage disease may be governed by a range of alternative pro-angiogenic factors, and there may be some redundancy for VEGF (Herbst et al. 2005; Relf et al. 1997).

Another tumor-specific molecule that seems essential for the survival of cancer is the epidermal growth factor receptor (EGFR). Arguably its deregulation involved in abnormal EGFR activity (Ciardiello and Tortora 2002) include: (a) autocrine activation by overproduction of

EGFR ligands, (b) EGF receptor overexpression, (c) another less well-understood mechanism is the constitutive activation of EGFR, and finally (d) defective internalization or down-regulation of the receptor.

To evaluate the role of EGFR in the outcome of patients with locally advanced and metastatic non-small cell lung cancer (NSCLC) in a subgroup of patients in the BR.21 study (Tsao et al. 2005), patients with advanced and relapsed NSCLC in second- and third-line (after failure of at least one chemotherapy) were randomized to erlotinib (Tarceva), a small molecule that inhibits the EGFR tyrosine kinase activity, or placebo ($n = 731$) (Shepherd et al. 2005). The primary endpoint was median overall survival, which was significantly longer with 6.7 months in patients receiving erlotinib compared to 4.7 months in patients receiving placebo. Retrospectively, selected patient samples were analyzed for mutations, EGFR overexpression in immunohistochemistry, and EGFR gene amplification by FISH analysis. In univariant analyses, survival was significantly prolonged in the erlotinib group compared to the placebo group when EGFR was overexpressed (HR 0.68; $p = 0.02$) or when high copy number was detected by FISH (HR 0.44: $p = 0.008$). In multivariant analysis, the survival after treatment with erlotinib was not dependent on EGFR overexpression, EGFR gene amplification, or EGFR mutations. However, higher patient numbers and prospective analysis will have to confirm these data.

Finally, the question of whether tumor targets always have to be measured and remain target-specific can be answered as follows:

- Not necessary to measure, e.g., VEGF for treatment with bevacizumab (Avastin) or CD20 for treatment with anti-CD20 monoclonal antibodies (Rituxan)
- Necessary to measure and provide relevant response prediction information, e.g., HER2 for trastuzumab (Herceptin), ER/PR for hormonal treatments
- Necessary to measure but currently insufficient information to provide definitive conclusions on response prediction, e.g., EGFR for erlotinib (Tarceva)

10.4 Dose and Dose Regimen Findings with Molecularly Targeted Drugs

The challenges in developing targeted agents are well documented, and considerable efforts have been invested in defining suitable molecular measures and target modulation to be associated with clinical benefit. Restricting patient enrollment to those with accessible disease for serial tumor sampling may decrease patient enrollment and add an ethical burden to the clinical protocol. The use of surrogate tissue (skin, mucosa, peripheral blood) or functional imaging technologies may provide appropriate solutions to these problems, assuming the changes in the surrogate tissue parallel those in the tumor. Finally, the optimal level of target inhibition must be determined and assays for measurement of the drug effect must be available (Pegram et al. 2005).

Phase I trials seek to determine the optimal or recommended phase II dose of a new compound for further testing (Parulekar and Eisenhauer 2004). For cytotoxic drugs, this dose corresponds to the highest dose associated with an acceptable level of toxicity and is derived from clinical data and preclinical dose–toxicity and dose–activity studies. The different mechanism of action and anti-tumor effect of the novel targeted compounds have been characterized by a lack of significant organ toxicity compared with conventional chemotherapy. The maximum tolerated dose may therefore greatly exceed the dose required to block target activity. The focus on optimum biological activity as a function of dose and regimen requires using appropriate biological markers as a potential surrogate for clinical endpoints. The optimum dose, however, may vary according to receptor saturation of different tumor types with different growth rates and individual factors influencing target expression or activity (Kelloff and Sigman 2005). Thus, although determination of the recommended phase II dose using toxicity as a surrogate endpoint for activity may not be necessary or unachievable in the phase I setting for these agents, the demonstration that the agents have the desired target effect is an important aspect of their early clinical development. We therefore may end up with two MTDs, the maximum tolerated dose and the maximum target inhibition dose on biomarkers of surrogate tissue.

Although the use of toxicity for dose selection may not be appropriate for agents that have maximal target inhibition at nontoxic doses, this method of dose selection may minimize the possibility that a subtherapeutic dose will be chosen for phase II. One example of this discrepancy is the comparison of the clinical outcome of two small molecules gefitinib (Iressa) and erlotinib (Tarceva), both inhibiting the EGFR tyrosine kinase. The ISEL study investigated the survival benefit of gefitinib at 250 mg/day (had shown no difference in previous phase II study for response rate compared to 500 mg/day) as a monotherapy in patients with advanced NSCLC who had failed one or two previous chemotherapies (Tamura and Fukuoka 2005). Approximately 1,700 patients were enrolled and randomized for gefitinib vs placebo. Although there was some improvement in survival with gefitinib vs placebo, the overall population failed to reach statistical significance in overall survival (HR 0.83, medium overall survival 5.6 months vs 5.1 months for placebo control). The BR.21 study was a phase III trial of erlotinib involving patients who showed progression after standard chemotherapy for non-small cell lung cancer in second or third line (Shepherd et al. 2005). Patients were randomly assigned in a 2:1 ratio to receive 150 mg of erlotinib daily or placebo. This compound, which showed higher sensitivity of different cell lines in in vitro cultures, was given at the maximum tolerated dose in the clinics. The response rate was 8.9% in the erlotinib group and less than 1% in the placebo group ($p < 0.001$). The median duration of the response was 7.9 months and 3.7 months for placebo group and the overall survival was significantly improved with 6.7 months vs 4.7 months (hazard ratio, 0.70; $p < 0.001$), in favor of erlotinib. These two compounds are good examples for the dichotomy of target modulation and clinical endpoints.

10.5 Appropriate Selection of Clinical Endpoints for Benefit Prediction

Conducting phase III trials involving targeted noncytotoxic molecules is no different from such trials testing cytotoxic agents since efficacy and safety remain the primary endpoints. As with conventional chemotherapy, phase III studies will ultimately define clinical utility because, regardless of the type of anticancer agent, the ultimate goal remains the same: prolongation of life and/or better relief of symptoms (Parulekar and Eisenhauer 2002). Several confounding factors interfere with the evaluation of survival benefit of new agents. Patients in the control arm may receive other investigational drugs, effective second- or third-line treatment becomes available for almost every tumor type, and quality-of-life assessment has to be integrated as an endpoint for relief of symptoms. Thus, recent focus on using biomarker endpoints as surrogates for survival will provide definitive estimates of clinical benefit in a shorter time (Kelloff and Sigman 2005).

Using progression-free survival (PFS) as an endpoint reflects more accurately the benefit of a new treatment, including tumor kinetics and its modulation under treatment, and removes confounding factors of subsequent therapies. Issues remain as to translating an improvement of PFS into clinical benefit and incorporating toxicity information and clinical benefit of tumor-related symptoms.

Given that many targeted treatments are noncytotoxic and therefore more difficult to assess than traditional cytotoxics, the transition to phase III requires a strategic decision based on new considerations, and response rate may not be an appropriate measure of anti-tumor activity.

10.6 Integration of New Drugs into Standard Treatment

Over the past decade, a number of new cytotoxic agents have become available for the treatment of cancer, including taxanes, gemcitabine and vinorelbine, and have especially been integrated into the treatment paradigm of non-small cell lung cancer. The combination of one or more of these agents with a platinum compound has resulted in an overall response rate of 20%, a time to tumor progression of 3.4–4.2 months (95% CI, 2.8–3.9 months) and a median overall survival of 7.5–8.0 months (95% CI, 7.0–9.5). No significant difference in survival was observed among the different combinations/regimens used (Schiller et al. 2002). As a result, several targeted treatments have been investigated in chemo-naïve patients with advanced and relapsed NSCLC in combination with chemotherapy. Agents such as inhibitors for EGFR tyrosine kinase, matrix metalloproteinases, farnesyl transferases, and protein kinase C were combined with paclitaxel and carboplatin and randomized vs chemotherapy alone. None of these phase III studies showed any significant survival benefit of the combination vs chemotherapy alone. These data may raise two hypotheses: (1) chemotherapy prohibits combination with targeted treatment in NSCLC or (2) chemotherapy induces confounding modulations of a tumor-specific target.

1. Chemotherapy prohibits combination with targeted treatment in NSCLC.

A study that addressed the first hypothesis of whether chemotherapy is prohibitive in combination with targeted treatment was investigated in a randomized phase II/III trial of paclitaxel plus carboplatin for six cycles with or without bevacizumab at 15 mg/kg every 3 weeks until tumor progression in patients with advanced non-squamous non-small cell lung cancer (ECOG 4599) (Sandler et al. 2005). In the second planned interim analysis, a significant improvement of objective response rate (10% vs 27%); progression-free survival (4.5 months vs 6.4 months) and overall survival (10.2 months vs 12.5 months) for the combination of chemotherapy with bevacizumab was observed. The treatment was well tolerated despite treatment-related deaths due to hemoptysis in five out of 357 patients treated with bevacizumab (Sandler et al. 2005). This was the first study to support the combination of targeted treatment in combination with chemotherapy in non-small cell lung cancer.

2. Chemotherapy induces confounding modulations of a tumor-specific target.

The question of whether chemotherapy induces a confounding modulation of the tumor-specific targets and therefore reduce the effect of targeted treatment was investigated in a trial with erlotinib plus gemcitabine vs gemcitabine alone in patients with advanced pancreatic cancer (Moore et al. 2005). A group of 569 patients were randomized to receive gemcitabine 1,000 mg/m^2 IV weekly for 7 of 8 weeks then weekly 3 out of 4 weeks plus either erlotinib 100 mg p.o. daily or a placebo. There was a significant improvement in overall survival that favored the combination with erlotinib (6.37 months) vs chemotherapy alone (5.91 months) with a hazard ratio of 0.81 (95% CI, 0.67–0.97). The corresponding 1-year survival rates were 24% vs 17% and the tumor control rates (CR/PR/SD) were 57% vs 49% for the erlotinib and placebo groups, respectively. Although a subgroup analysis of EGFR-positive tumor suggested a benefit for patients with erlotinib, prospective studies in larger patient groups will have to confirm these data. These results demonstrate for the first time a benefit of the combination of a tumor-specific agent with chemotherapy.

10.7 Conclusions

Improved understanding of cancer biology has provided many new targets for cancer treatment. Several targets have been validated, although overall clinical progress remains slow. Targeted drug development with dose and dose findings remains complex, as patient selection and appropriate pharmacodynamic endpoints are critical. However, targeted drugs have improved outcomes to an extent not seen before.

References

Baselga J (2001) Herceptin alone or in combination with chemotherapy in the treatment of HER2-positive metastatic breast cancer: pivotal trials. Oncology 61:14–21

Ciardiello F, Tortora G (2002) Anti-epidermal growth factor receptor drugs in cancer therapy. Expert Opin Investig Drugs 11:755–768

Ciardiello F, Tortora G (2003) Epidermal growth factor receptor (EGFR) as a target in cancer therapy. Eur J Cancer 39:1348–1354

Cobleigh M, Vogel C, Tripathy D, Robert N et al (1999) Multinational study of the efficacy and safety of humanized anti-HER2 monoclonal antibody in women who have HER-2 overexpressing metastatic breast cancer that has progressed after chemotherapy for metastatic disease. J Clin Oncol 17:2639–2648

Cortes J, Baselga J, Wardley A, Bianchi G et al (2005) Open-label, randomized, phase II study of pertuzumab (Omnitarg) in patients with metastatic breast cancer with low expression of HER2. Proc Am Soc Clin Oncol 23: abstr. 5051

Ferrara N, Gerber HP, LeCouter J (2003) The biology of VEGF and its receptors. Nat Med 9:669–676

Herbst R, Onn A, Sandler A (2005) Angiogenesis and lung cancer: prognostic and therapeutic implications. J Clin Oncol 23:3243–3256

Hurwitz H, Fehrenbacher L, Novotny W, Cartwright T et al (2004) Bevacizumab plus irinotecan, fluorouracil, and leucovorin for metastatic colorectal cancer. New Engl J Med 350:2335–2342

Kelloff G, Sigman C (2005) New science-based endpoints to accelerate drug development Eur J Cancer 41:491–501

Mass R, Sanders C, Charlene K, Everett T, Johnson L (2000) The concordance between the clinical trials assay (CTA) and fluorescence in situ hybridization (FISH) in Herceptin pivotal trials. Proc Am Soc Clin Oncol 19: abstr. 75A

Miller K, Chap L, Holmes F, Cobleigh M et al (2005) Randomized phase III trial of capecitabine compared with bevacizumab plus capecitabine in patients with previously treated metastatic breast cancer. J Clin Oncol 23:792–799

Moore M, Goldstein D, Hamm J, Figer A (2005) Erlotinib plus gemcitabine compared to gemcitabine alone in patients with advanced pancreatic cancer. Proc Am Soc Clin Oncol 23: abstr. 1

Nahta R, Hung MC, Esteva F (2004) The HER-2-targeting antibodies trastuzumab and pertuzumab synergistically inhibit the survival of cancer cells. Cancer Res 64:2243–2246

Parulekar W, Eisenhauer E (2002) Novel endpoints and design of early clinical trials. Ann Oncol Suppl 4:139–143

Parulekar W, Eisenhauer E (2004) Phase I trial design for solid tumor studies of targeted, non-cytotoxic agents: theory and practice. J Natl Cancer Inst 96:990–997

Pegram M, Pietras R, Bajamonde A, Klein P, Fyfe G (2005) Targeted therapy: wave of the future. J Clin Oncol 23:1776–1781

Relf M, Lejeune S, Scott P et al (1997) Expression of the angiogenic factors vascular endothelial cell growth factor, acid and basic fibroblast growth factor beta-1, platelet-derived endothelial cell growth factor, placenta growth factor, and pleiotrophin in human primary breast cancer and its relation to angiogenesis. Cancer Res 57:963–969

Roberts T, Lynch T, Chabner B (2003) The phase III trial in the era of targeted therapy: unraveling the "go or no go" decision. J Clin Oncol 21:3683–3695

Sandler A, Gray R, Brahmer J, Dowlati A et al (2005) Randomized phase II/III trial of paclitaxel plus carboplatin with or without bevacizumab in patients with advanced non-squamous non-small cell lung cancer. Proc Am Soc Clin Oncol 24: abstr. LBA4

Schiller J, Harrington D, Belani C, Langer C et al (2002) Comparison of four chemotherapy regimens for advanced non-small-cell lung cancer. New Engl J Med 346:92–98

Shepherd F, Pereira J, Ciuleanu T, Tan E et al (2005) Erlotinib in previously treated non-small-cell lung cancer. New Engl J Med 353:123–132

Slamon D, Leyland-Joanes B, Shak S, Fuchs H et al (2001) Use of chemotherapy plus a monoclonal antibody against HER2 for metastatic breast cancer that overexpresses HER2. New Engl J Med 344:783–792

Tamura K, Fukuoka M (2005) Gefitinib in non-small cell lung cancer. Expert Opin Pharmacoth 6:985–993
Tsao M, Sakurada A, Cutz JC, Zhu C et al (2005) Erlotinib in lung cancer – molecular and clinical predictors of outcome. New Engl J Med 353:133–144
Vogel C, Cobleigh M, Tripathy D, Gutheil J et al (2002) Efficacy and safety of Trastuzumab as a single agent in first-line treatment of HER2-overexpressing metastatic breast cancer. J Clin Oncol 20:719–726

11 Exploratory IND: A New Regulatory Strategy for Early Clinical Drug Development in the United States

N. Sarapa

Abstract. Optimizing exploratory drug development by means of doing first-in-human studies earlier is an attractive option for pharmaceutical sponsors to select more successful drug candidates. Traditional registration-driven clinical phase 1 programs could be preceded by early human screening studies with subpharmacological single doses (microdoses) or with low pharmacologically active doses of one or several lead candidates, whereby human pharmacokinetic and pharmacodynamic data are generated very early. Such low doses are not expected to have clinically significant toxic potential, so early human screening studies may be supported by abbreviated nonclinical safety packages. Recent US FDA draft guidance (April 2005) regulated early human screening studies conducted under the exploratory IND. The author outlines the features of the study design, dose selection, and nonclinical safety packages required in support of exploratory IND studies in humans. In appropriately chosen cases, exploratory IND could allow for patients' quicker access to safer and more efficacious doses of novel drugs, reduce attrition in clinical trials, and facilitate more economical drug development.

11.1 Introduction: Challenges in Drug Development Today

Patients, physicians, and pharmaceutical sponsors alike want new, more effective and safer medicines faster, simultaneously with an improved cost-effectiveness of drug development. Drug discovery, lead candidate selection, and nonclinical development are undergoing rapid changes that are driven, in part, by scientific advances in areas such as combinatorial chemistry, molecular and cell biology, and high-throughput technology, and, in part, by strong competition and economic forces (Lesko et al. 2000). As a result, the pressure to accelerate drug discovery and development is increasing, especially in the phases leading up to and during early human clinical testing. The lead candidate selection process must focus on greater predictivity of success in later-phase clinical trials, so that better therapeutic agents with a lower risk are identified earlier, thereby resulting in safer, more effective, and more economical drug development programs.

In March 2004, the US Food and Drug Administration (FDA) issued a seminal position paper entitled "Innovation or stagnation? Challenge and opportunity on the critical path to new medical products," pointing out the major problems and challenges in drug development today (Food and Drug Administration 2004). The drug discovery and approval process is long, expensive, and inefficient. During the last several years, the number of new drug and biologic applications submitted to the FDA has declined significantly; the number of innovative medical device applications has also decreased. In contrast to such decreased pharmaceutical R&D output, the costs of product development have soared by 55% during the last 5 years. The current cost of bringing a new medicine to market, estimated to be between US $0.8 and 1.7 billion, is a major burden for developing innovative, higher-risk drugs. The very high attrition rate of drug candidates in clinical development is another critical issue. Historically, 14% of products entering clinical testing had gained regulatory approval and entered the market, but today the chance of a candidate drug entering Phase 1 trials being approved is only 8%. This is an alarming statistic for pharmaceutical industry, and traditional drug development paradigms are not sustainable in the face of current and future drug development challenges. In the FDA's view, the sciences applied to clinical product development have not kept pace with the tremendous advances in the basic sciences. In many cases, drug developers are using the tools and concepts of the last century to assess this century's candidates. The FDA's Critical Path paper submits that a new product development toolkit containing powerful new scientific and technical methods such as animal or computer-based predictive models, biomarkers for safety and effectiveness, and new clinical evaluation strategies, is urgently needed to improve predictability and efficiency along the drug development path from laboratory concept to commercial product.

In order to increase pipeline flow, pharmaceutical sponsors should either invest much more in R&D or find a new way to bring more promising candidates into the clinic. Many sponsors cannot invest in R&D on a scale typical of a very large pharmaceutical company, so instead they should be more innovative, flexible, and efficient to increase productivity and improve competitiveness as global R&D-based innovator pharmaceutical companies.

Pharmaceutical companies recognize the need to identify the compounds with deficient properties long before their entry into Phase 1. As a consequence, drug discovery scientists now routinely screen potential clinical candidates using a variety of *in vitro*, *ex vivo* and *in silico* technologies, as well as the more traditional *in vivo* animal models. The predictive power of preclinical drug metabolism and pharmacology models has advanced considerably, due to an increased understanding of the relationships between *in vitro*, animal, and human pharmacokinetics (PK) and pharmacodynamics (PD). The focused application of preclinical-to-clinical extrapolation based on PK and PD has improved the efficiency of drug development in the pharmaceutical industry (Reigner et al. 1997). However, there are still cases where drug disposition and therapeutically effective plasma concentrations in humans are not readily predictable from preclinical data. Up to 40% of compounds entering clinical development fail because of inappropriate drug metabolism and PK (Lappin et al. 2004). As reported by Frank and Hargreaves (2003), the major reasons for attrition in the clinic from 1991 to 2000 were lack of efficacy (accounting for approximately 30% of failures) and safety problems (accounting for approximately 30% more). It is quite possible that many of the efficacy and safety failures are related to PK and PK/PD issues. A major factor that could boost success rates in the clinic would be better characterization of drug candidates at the preclinical/early clinical interface by obtaining human data earlier. The traditional registration-driven Phase 1 trials could in certain cases be preceded by innovative science-driven early first-in-human (FIH) screening studies with a single subpharmacological dose (microdose) or with low pharmacologically active doses of one or several new chemical entities (NCEs). This approach, with appropriate safeguards, might accelerate drug development without lowering clinical safety standards and allow for better and more informed decision making in less time.

11.2 New Regulatory Pathways for Exploratory Development

11.2.1 European Position Paper on Human Microdosing

The regulatory authorities worldwide increasingly advocate novel paradigms that offer potential to accelerate patients' access to more optimized new drugs. In 2004, the Committee for Human Medicinal Products (CHMP) of the European Medicines Evaluation Agency (EMEA) adopted a position paper on the nonclinical safety studies to support clinical trials with a single microdose [6]. It describes screening FIH trials that are exploratory by nature and are conducted before Phase 1 (in the so-called phase 0) with one or several closely related drug candidates. The EMEA position paper defines the human microdose as the 1/100th of the dose calculated to yield a pharmacological effect of the compound based on primary pharmacodynamic data obtained from animal and in vitro models (typically this would be in the low microgram range) and prescribes that total amount of compound(s) administered in a human microdosing study must not exceed 100 μg (European Agency for the Evaluation of Medicines for Human Use 2004). Although the maximum dose of 100 μg in the EMEA position paper is debatable, this document provides a useful and streamlined regulatory pathway in the European Union for early selection between drug candidates by human screening PK studies. Importantly, EMEA endorsed the abbreviated nonclinical safety package for human microdosing studies that is purposefully less comprehensive than the ICH M3 guidance applicable to traditional Phase 1 studies (International Conference on Harmonization 2000). The EMEA's enlightened and visionary approach shifted the focus of early drug development away from the less than predictive animal models toward conducting safe, ethical, and timely studies in humans – the actual target species. This approach recognized that the best model for humans is indeed humans themselves.

11.2.2 US FDA Exploratory IND Guidance

In the US, the approval for first-in-human study by the US Food and Drug Administration (FDA) requires prior submission of an application for an investigational new drug (IND). In general, there are three types (investigator IND, emergency use IND, and treatment IND), and two categories (commercial and research) of IND. Within the commercial IND category, pharmaceutical sponsors can pursue full (traditional) INDs and expedited (abbreviated) INDs. Recently, a new category of expedited IND has been established in the US, namely the exploratory IND. In April 2005, the FDA issued a draft guidance document for exploratory investigational new drug applications that describes very early Phase 1 exploratory approaches (microdose and low pharmacological dose) and enables pharmaceutical sponsors to better develop promising drug candidates while protecting the well-being of human study subjects (Food and Drug Administration 2005).

The concept of an exploratory IND was proposed to the FDA in 2004 by the American drug industry's professional association, the Pharmaceutical Research and Manufacturing Association (PhRMA) as a much improved version of the very rarely used FDA screening IND Manual of Policies and Procedures (MAPP). The purpose of an exploratory IND is to allow for early clinical testing of one or several new chemical entities based on a reduced preclinical package (CMC and animal data). Once a suitable candidate is identified for further development, a full traditional IND has to be submitted. According to the draft FDA guidance, approval may be given to an exploratory IND provided that the sponsor:

1. Provides an acceptable initial safety profile (as described above)
2. Opens preliminary discussion with the agency on detail study plans
3. Defends proposed study design and rationale

Study Design

For the purposes of the FDA guidance, the term "exploratory IND study" describes a clinical trial that occurs very early in Phase 1, involves very limited human exposure, and has no therapeutic intent (e.g., screening studies, microdose studies). Likewise, the exploratory IND studies involve dosing a limited number of subjects with a limited dose range for a limited period of time. Such exploratory IND studies are conducted prior to the traditional dose escalation, safety, and tolerance studies that ordinarily initiate a full Phase 1 clinical drug development program. The duration of dosing in an exploratory IND study is expected to be limited (e.g., 7 days).

The FDA recommended the following types of human studies that could be proposed for conduct under the exploratory IND:

- Single-dose studies
  - Microdose studies with subpharmacologic doses or studies with low pharmacological doses investigating mostly pharmacokinetic endpoints. For example, microdosing, pharmacokinetic, and biodistribution studies
- Multiple-dose studies
  - Investigating pharmacokinetic or pharmacodynamic endpoints as appropriate
  - Limited dosing duration (e.g., $\sim$ 7 days)
  - If dose escalating studies are proposed, they must investigate a pharmacodynamic endpoint

Note that these clinical study designs are made deliberately flexible to allow for identification of critical data for early go/no-go decisions in clinical development. The data from these studies can and will be acceptable to support later submissions of a traditional IND.

Microdosing Under the Exploratory IND

The FDA definition of a microdose is identical to that in the EMEA document, and the toxicology requirements for these trials are very similar. A single mammalian species can be used if justified by *in vitro* metabolism data and by comparative data on *in vitro* pharmacodynamic effects [8]. The route of exposure in animals should be the intended clinical route. Animals in the toxicology study should be observed for 14 days after dosing with an interim sacrifice, typically on day 2, and endpoints evaluated should include body weights, clinical signs, clinical chemistries, hematology, and histopathology. The study should be designed to establish a dose inducing a minimal toxic effect, or alternatively, establishing a margin of safety. To establish a margin of safety, the sponsor should demonstrate that a large multiple (e.g., 100×) of the proposed human dose does not induce adverse effects in experimental animals. Scaling from animals to humans based on body surface area can be used to select the dose for use in the clinical trial. Because microdose studies involve only single exposures to microgram quantities of test materials and because such exposures are comparable to routine environmental exposures, genetic toxicology testing is not needed.

The exploratory IND will allow human microdosing studies to be undertaken in the US as easily as is currently done in Europe. Moreover, the experience from certain sponsors indicates that the FDA may be willing to accept a non-GMP drug substance for human use in a microdosing study, as long as the sponsors submit evidence of detailed characterization of an active pharmaceutical ingredient.

Low Pharmacological Dosing Under the Exploratory IND

The following principles should be respected when selecting the starting and maximum human doses for exploratory IND studies utilizing low pharmacologically active doses. The starting human dose is anticipated to be no greater than 1/50 of the no-observable-adverse-effect-level dose (NOAEL) from the 2-week toxicology study in the sensitive species on a milligram per square meter basis. The maximum human dose would be the lowest of the following:

1. One-quarter of the NOAEL in a 2-week toxicology study

2. One-half of the AUC at the NOAEL in the 2-week rodent study, or the AUC in the dog at the rat NOAEL, whichever is lower, or
3. The dose that produces a pharmacological response or at which target modulation is observed in the clinical trial

Escalation from the proposed stopping dose should only be performed after consultation and in agreement with the FDA. Note that clinical testing within exploratory IND will never be targeted at maximum tolerated dose (MTD) in humans, as is routine in classic FIH studies conducted under the traditional INDs.

Nonclinical Safety Package in Support of Low Pharmacological Dosing The scope of the nonclinical safety program for an exploratory IND is intended to be less comprehensive than that recommended by ICH M3 Guidance (International Conference on Harmonization 2000) and the FDA traditional IND package for clinical studies utilizing higher doses or exposures. The requirements for nonclinical testing in support of an exploratory IND study were defined as commensurate with the limited clinical risks induced by lower doses and shorter duration of exposure. This means that the nonclinical safety package is focused on generating adequate information to assure clinical safety for the initial clinical dose up to the planned (low) maximal dose for the (limited) duration of the clinical trial proposed under the exploratory IND.

The abridged nonclinical safety package for exploratory IND will consider the following:

- The pharmacological activity of NCE may be primary or secondary
- Metabolism in toxicology species must be proven as comparable to man (*in vitro*)
- Any known species sensitivity must be taken into consideration
- One *in vitro* genotoxicity study is needed
- Toxicokinetics is mandatory to prove adequate exposure in toxicology species
- A combination of GLP and robust non-GLP studies is permissible

Toxicology The following abbreviated toxicology evaluation is required to support the exploratory IND study with a low pharmacologically active dose:

- Repeat-dose clinical trials lasting up to 7 days are supported by a 2-week repeat dose toxicology study (usually in rat) with clinical chemistry, toxicokinetics, and histopathology microscopic assessment of major organs. (For example, three males, three females, IV, and the intended route of human exposure; 50-fold safety factor)
  - Must be done according to GLP
  - Identify organ toxicity and NOAEL
- Sensitivity confirmation in a second species (usually dog; dose ranging or repeat administration at rat NOAEL on equivalent occasions as the intended regimen in humans)
  - Toxicokinetics, intended clinical route

Safety Pharmacology In preparing the safety pharmacology package for exploratory IND, major systems to be tested include cardiovascular in non-rodent species (required), central nervous, and respiratory systems. The latter two systems can be assessed in rodent along with the general toxicology program package. However, the cardiovascular safety pharmacology must be done in non-rodent with the following criteria.

- It may be incorporated into species-sensitivity assay.
- It need not exceed the exposure at rodent NOAEL.
- When the stand alone approach is used, the animals should be monitored for toxic effects and as feasible for clinical events, clinical chemistry, and hematology results.

A cardiovascular safety study need not provide histopathology, and unlike the ICH Guideline on safety (S7A), testing to toxic levels is not required. Only in the case of moribund or unscheduled deaths would gross or microscopic examination of organs from the cardiovascular safety study be expected.

Genetic Toxicology In general, each product in this type of exploratory IND should be tested for potential genotoxicity unless such testing is not appropriate for the population to be studied. The genetic toxicology tests should include:

- Bacterial mutation assay (all strains and exposures)
- Chromosomal aberrations (*in vitro* or *in vivo*)
- Two *in vitro* or one *in vitro* plus one *in vivo* assay to evaluate mutagenicity and clastogenicity:
  - Can use bone marrow from repeat dose rodent study for *in vivo*
  - Screening assays with *limited study designs* are acceptable (e.g., a limited number of Ames strains)
  - May not be full GLP

Pharmacokinetics The pharmacokinetic (ADME) data submitted in the exploratory IND should include:

- Standard *in vitro* assays
  - Solubility
  - Permeability
  - Metabolic stability
  - Protein binding
- *In vivo* oral and IV pharmacokinetics (one or two species)
- Comparable metabolism between toxicological species and humans *in vitro*
- Toxicokinetic (TK) data from the pivotal toxicological study
- Preclinical proof of PK linearity (pharmacological dose PK in non-rodent species)

Chemistry, Manufacturing, and Quality Control The information on Chemistry, Manufacturing, and Quality Control (CMC) of an NCE is required for obtaining approval for exploratory IND, but the reduced CMC package recommended by the FDA (Food and Drug Administration 2005) includes some important allowances for flexibility in GMP compliance. The active pharmaceutical ingredient (API) for exploratory IND does not have to be characterized according to full GMP principles. A non-GMP API for exploratory IND may be acceptable if:

- Produced in a medicinal or process laboratory
- Good laboratory notebook documentation to support CMC
- Adequate structural and purity characterization of final material

- The simplest formulation is used, e.g., extemporary preparation of the drug in a bottle
- Possess limited stability testing
- Some QA involvement of release process

In describing the drug substance, the following descriptors are recommended for exploratory IND:

- Description (physical, chemical, or biological, as appropriate)
- Name and address of manufacturer
- General method of preparation
- Acceptance limits and methods used to ensure identity and purity
- Stability data to support proposed duration of clinical studies

The analytical methods and characterization to be used for the drug substance in exploratory IND are expected to include descriptions under the following headings.

- Identity, structure, purity assay, potency assay (biologics) impurity profile, residual solvents, and heavy metals and optical rotation (as appropriate).
- When API for toxicology are from the *same batch* as the API for the clinical study, there is no need to demonstrate representativeness.
- Characterization of impurity profiles is not considered necessary when purity is greater than 95%; optical rotation is not needed.

When the API for clinical and toxicology studies are from different batches, representative API needs to be demonstrated using appropriate tests, as described in the FDA CMC guidance for the traditional IND.

In describing the drug product, the following descriptors are recommended as appropriate.

- List of components
- Quantitative composition of product
- Brief general description of method of manufacture and packaging
- Acceptance limits and analytical limits used to ensure identity, strength, quality, and purity
- Stability support data

11.3 Conclusion

Early human screening studies utilizing microdoses or low pharmacologically active doses are not appropriate for every drug development program. However, in carefully selected cases, a first-in-human screening study can be conducted early and in a medically safe and ethical manner based on abbreviated nonclinical safety packages described in the FDA's guidance on the exploratory IND. If applied selectively, the exploratory IND approach could improve the quality of internal decision making by sponsors based on exploratory human ADME/PK or PD data obtained early, i.e., before the investment is made in substantial resources needed for traditional Phase 1 clinical development. This approach can improve the chances of selecting the best candidate and enable allocation of resources toward the most deserving drug candidates, resulting in saving time, money, and resources during preclinical and clinical development. Because the chosen candidate will have favorable PK and PD properties, the risk of human exposure to NCEs with poorly predictable effects would be minimized. Early knowledge of human PK and PD might enable dose escalation schedules to be optimized, resulting in fewer healthy subjects being exposed in Phase 1 studies before doses and dose regimens for Phase 2 are selected with confidence.

Against the backdrop of obvious pressure to bring more therapeutic drugs to market in the area of unmet medical needs, the increasing public demands for affordable, safe, and more efficient drugs, and the FDA's own Critical Path initiatives, the exploratory IND path might be more frequently used by pharmaceutical sponsors keen to improve the success rate in later phases of clinical drug development. By issuing the exploratory IND guidance, the FDA showed commitment to their Critical Path initiative by means of facilitating the process of identifying human PK and PD properties before commencing traditional Phase 1 trials. By virtue of improving the selection of NCEs, early human screening studies under the Exploratory IND would provide patients with quicker access to safer and more effective drugs and simultaneously identify failures earlier, leading to more efficient drug development.

References

European Agency for the Evaluation of Medicines for Human Use (EMEA) (2004) Position paper on nonclinical safety studies to support clinical trials with a single microdose. CHMP/SWP/2599/02/Rev 1, London

Food and Drug Administration (2004) Innovation or stagnation? Challenge and opportunity on the critical path to new medical products. http://www.fda.gov/oc/initiatives/criticalpath/whitepaper.html. Cited 6 April 2006

Food and Drug Administration (2005) Draft guidance for industry, investigators, and reviewers. Exploratory IND studies. Rockville

Frank R, Hargreaves R (2003) Clinical biomarkers in drug discovery and development. Nature Rev Drug Discov 2:566–580

International Conference on Harmonization (2000) Harmonized Tripartite Guideline (M3). Nonclinical safety studies for the conduct of human clinical trials for pharmaceuticals, issued 16 July 1997 and amended 9 November 2000

Lappin G, Garner RC (2003) Big physics, small doses: the use of AMS and PET in human microdosing of development drugs. Nature Rev Drug Discov 7:233–240

Lesko LJ, Rowland M, Peck CC, Blaschke TF (2000) Optimizing the science of drug development: opportunities for better candidate selection and accelerated evaluation in humans. Pharmaceut Res 17:1335

Reigner BG, Williams PEO, Patel IH, Steimer JL, Peck C, van Brummelen P (1997) An evaluation of the integration of pharmacokinetic and pharmacodynamic principles in clinical drug development. Experience within Hoffmann La Roche. Clin Pharmacokinet 33(2):145–152

12 Ethnic Aspects of Cancer Trials in Asia

T.W.T. Leung

Abstract. New drugs which have potential in cancer therapy are emerging every day and there is an increasing demand for trial patients all over the world. Asia being the most populated continent, as well as a large market for drug sales, has its own epidemiology of disease. The Asian races also have specific genomics that might affect response to drug treatment. In addition, there are cultural issues to be considered when considering clinical trials. In conclusion, more clinical trials should be done among Asian populations for the best results of drug treatment.

Today, cancer still remains as one of the most important health hazards in the world. Progress in terms of cure has been slow due to complexity of the disease. Current available treatment modalities, particularly surgery and radiotherapy, are only effective when the cancer is discovered in early stages. Treatment for advanced-stage cancer, which is often palliative, largely depends on drug therapy. Drug therapy is also important and increasingly so as an adjunct treatment to curative surgery or radiotherapy. However, available drugs are still limited because drug development is a lengthy process and the majority of investigative agents could not pass either preclinical or clinical testing. Discovery of new drugs continues to be the most important task in the battle against cancer.

With the rapid explosion of knowledge in biotechnology and pathogenesis of cancer in the last two decades, discovery and synthesis of potentially useful cancer drugs has become more efficient. However, extensive testing in humans requires large numbers of subjects for clinical trial before the drug can be approved for treatment. Therefore, the demand for patient accrual for cancer drug trials is high.

Asia has the largest population of all continents. Socioeconomic development in Asia has been rapid over the last two decades, which makes Asia now the largest market for drug sales. In the past, most of the drugs used in Asia follow the recommendations of Western countries. The regulatory authorities in North America and Europe approve drugs based on clinical trials that were conducted in non-Asian countries. It was assumed that the efficacy and adverse reaction of new drugs should be similar in Asian and non-Asian populations, but our experience in cancer trials has suggested the contrary.

12.1 Ethnic Differences in Cancer Epidemiology

The types of cancer that we see in Asian countries are very different from those of Western countries. Cancers such as hepatocellular carcinoma are rarely seen in Western countries and nasopharyngeal carcinoma is a unique disease of southern Chinese populations. These are all common cancers that we see in Asia, but there is no treatment paradigm that we can derive from clinical trials conducted in non-Asian countries for these cancers.

Hong Kong, mainly populated by Chinese, has a slightly different cancer epidemiology compared with non-Asian populations. From the Hong Kong Cancer registry report in 2002, the most common cancers are lung, colon, and breast cancer, which is no different from Western countries. However, the incidence rate of hepatocellular carcinoma in Hong Kong was 23.2, much higher than the 2.4 per 100,000 population in the United States (Hong Kong Cancer Registry Report 2002; El-Serag and Mason A 1999). Nasopharyngeal carcinoma is also unique for Southern Chinese: the incidence rate was 14.2 per 100,000 in Hong Kong populations (Hong Kong Cancer Registry Report 2002). This disease is virtually unseen in non-Asian populations. Going to other parts of Asia, the cancer epidemiology of Taiwan and Singapore are similar to Hong Kong because they are mainly populated by Chinese. Japan and Korea, however, have a high incidence of stomach cancer compared to the rest of the world. This paper does not intend to go into details of cancer epidemiology in different Asian countries but points out that there is an ethnic difference in cancer epidemiology. The implication for cancer clinical trials is that rapid accrual of patients can be achieved for common cancers in a particular ethnic group. Larger impact on healthcare is also expected when a new treatment is found for these common cancers.

12.2 Differences in Pathology and Etiology of Cancer

Though similar in histology, certain types of cancer have different etiologies and clinical behaviors. Hepatocellular carcinoma is common both in Japan and Hong Kong. However, the viral etiology of hepatocellular carcinoma is different between these two ethnic groups. In Hong Kong, most hepatocellular carcinoma is associated with hepatitis B (80.3%) and less for hepatitis C (7.3%) (Leung and Tam 1992). In Japan, only 14% of hepatocellular carcinoma is associated with hepatitis B but 81% with hepatitis C (Kiyosawa and Umemura 2004). This is also similar to other non-Asian populations. Differences in viral etiology might have different clinical behaviors of disease and response to drug treatment. Hepatitis C-associated hepatocellular carcinoma commonly has more multifocal disease and severe liver fibrosis or cirrhosis. The response to chemotherapy is also different (Wong and Chan 2005). Similarly, there are more nonsmoking related female lung cancers in Chinese populations than those in Western countries (Lam and White 2004). There is also a higher incidence of adenocarcinoma among Chinese women having lung cancer. This accounts for the difference in outcome sometimes seen in clinical trials conducted in Asia and non-Asian populations. It is also important that ethnic groups and etiology of cancer should be stratified in global clinical trials.

12.3 Interethnic Differences in Drug Response

We are now in the genomic era of medicine. Differences in drug response and clearance are related to the genetic makeup of individual subjects. Among the drug metabolizing enzymes that show pharmacogenetic variation between individuals, at least 77% show an interethnic difference of variant frequency (Kalow 2002). We have conducted a study on Chinese breast cancer patients who received a combination of doxorubicin and cyclophosphamide chemotherapy and found that Chinese patients had worse bone marrow toxicity than Western populations receiving similar treatment (Ma and Yeo 2002). From a phase II clinical trial treating lung cancer patients with a combination of docetaxel and carboplatin, it was found that Asian patients had a better response rate but more neutropenic episodes than non-Asians (Millward and Boyer 2003). This might be due to the specific genotype of Asians, which is associated with a delay in docetaxel clearance (Goh and Lee SC 2002). Therefore interethnic differences in drug response should be observed and can be crucial in cancer trials.

12.4 Cultural Differences

There are also cultural differences in cancer trials between Asian and non-Asian populations. Clinical trials are widely accepted by healthcare workers and patients in non-Asian populations. However, only the more developed countries in Asia such as Singapore, Japan, Taiwan, and a few cities in China such as Beijing, Shanghai, and Hong Kong can carry out clinical trials according to good clinical practice. For the other countries, patients still expect their doctors to give them the best treatment rather than entering them into clinical trials. This can make accrual of trial subjects very difficult. It is also common in less developed areas that medical practice is less evidence-based. Patients tend to use a lot of alternative medicines and might not wish to receive protocolized treatment.

12.5 Conclusion

Ethnic differences in drug response should be recognized and is important in designing a good clinical trial. On the other hand, recognition of differences in disease characteristics aids in better stratification of trial patients. Lastly, cultural differences might affect the accrual rate, drop out rate, and credibility of data when trials are conducted in less developed countries.

References

El-Serag HB, Mason AC (1999) Rising incidence of hepatocellular carcinoma in the United States. N Eng J Med 340:745–750

Goh BC, Lee SC (2002) Explaining interindividual variability of docetaxel pharmacokinetics and pharmacodynamics in Asians through phenotyping and genotyping strategies. J Clin Oncol 20:3683–3690

Hong Kong Cancer Registry Report (2002) Hong Kong Hospital Authority, Hong Kong

Kalow W (2002) Interethnic differences in drug response. In: Kalow W (ed) Pharmacogenomics. Marcel Dekker, New York

Kiyosawa K, Umemura T (2004) Hepatocellular carcinoma: recent trends in Japan. Gastroenterology 127 [5 Suppl 1]:S17–S26

Lam WK, White NW (2004) Lung cancer epidemiology and risk factors in Asia and Africa. Int J Tuberc Lung Dis 8:1045–1057

Leung N, Tam J (1992) Does hepatitis C virus infection contribute to hepatocellular carcinoma in Hong Kong. Cancer70:40–44

Ma B, Yeo W (2002) Acute toxicity of adjuvant doxorubicin and cyclophosphamide for early breast cancer – a retrospective review of Chinese patients and comparison with an historic Western series. Radiother Oncol 62:185–189

Millward MJ, Boyer MJ (2003) Docetaxel and carboplatin is an active regimen in advanced non-small-cell lung cancer: a phase II study in Caucasian and Asian patients. Ann Oncol 14:449–454

Wong N, Chan KY (2005) Transcriptional profiling identifies gene expression changes associated with IFN-alpha tolerance in hepatitis C-related hepatocellular carcinoma cells. Clin Cancer Res 11:1319–1326

13 Evaluation of the Effect on Cardiac Repolarization (QTc Interval) of Oncologic Drugs

J. Morganroth

Abstract. The 12-lead electrocardiograph (ECG) is the standard safety measurement used in clinical trials to identify drug-induced cardiac adverse effects. Drug-induced prolongation of the QTc interval (the measure of cardiac repolarization change), when excessive and in conjunction with the right risk factors, can degenerate into a polymorphic ventricular tachycardia called torsades de pointes and has become a new focus for new drug development. The assessment of an

ECG in clinical practice using machine-defined QTc duration is intrinsically unreliable. Current regulatory concepts have focused on the need for measuring ECG intervals using manual techniques using digital processing in a central ECG laboratory. The QT interval is subject to a large degree of spontaneous variability requiring attention to basic clinical trial design issues such as sample size (use as large a cohort as possible), frequency of measurements taken (at least three to six ECGs at baseline and at many time points on therapy with pharmacokinetic samples if possible), and their accuracy. Since most oncologic products are cytotoxic, a Thorough or Dedicated ECG Trial cannot be conducted and in the usual trail, especially in phase I, all changes seen on the ECG will be attributed to the new oncology drug. For most nononcologic drugs, there is regulatory guidance on how much an effect on QTc duration might be related to the risk of cardiac toxicity. For oncology products, the central tendency magnitude and proportion of outliers needs to be well defined to construct a label if the risk–benefit analysis leads to marketing approval. Clinical cardiac findings such as syncope, ventricular tachyarrhythmias, and other cardiac effects will be important in this analysis.

13.1 Background

The cardiac adverse effects of oncologic therapies have been long recognized as one of the major organ toxicities that limit therapy for malignancies. The focus for such effects has historically been concentrated on cardiac contractile depression, myocardial ischemia, alterations in blood pressure, myocarditis, cardiac tamponade, hemorrhagic myocarditis, endomyocardial fibrosis, and bradyarrhythmias. These have been recently reviewed in detail and will not be covered in this report (Yeh et al. 2004). Instead, we will concentrate on the less well known and appreciated effect that oncologic agents may effect cardiac repolarization.

Drug-induced prolongation of the QTc interval (the measure of cardiac repolarization change), when excessive and in conjunction with the right risk factors can degenerate into a polymorphic ventricular tachycardia called torsades de pointes (TdP). While it is clear that the QT interval on the ECG may not be the best index of cardiac repolarization and its consequences, changes in the QTc duration is the one relied upon by regulatory authorities as the predictor of a new drug's cardiac safety. The commonest cause of new drug development delays, disapprovals, or removal from the market since the 1990s is the occurrence of drug-induced lengthening of the QTc interval. Examples of drugs with this fate have come from many different therapeutic groups and from many different, related and unrelated chemical structures. Examples include the antihistamine terfenadine, the antibiotic grepafloxacin, the antispasmodic terodiline, the neuroleptic droperidol, the atypical antipsychotic sertindole, the opioid levomethadyl, and the prokinetc agent cisapride.

Thus, the effect of new therapies on the ECG has taken special attention since the mid to late 1990s.

13.2 Basic ECG Issues Relevant to Clinical Research

The 12-lead ECG is a common standard safety measurement used in clinical trials to identify drug-induced cardiac adverse effects such as:

- Rhythm abnormalities (bradycardia, tachycardia, atrial fibrillation)
- New conduction abnormalities (atrioventricular or intraventricular block)
- Changes in ST-T-U morphology
- Evidence of myocardial injury or infarction, and now with special emphasis on:
  - Changes in repolarization interval (QTc interval)

Historically, in clinical trials, sponsors have relied upon the individual sites to detect drug-induced ECG changes and record their findings in the case report form. The assessment of an ECG in clinical practice can be made by either the clinician or by automated ECG cart software systems, which print the results on the top of the paper ECG. However, neither method is intrinsically reliable and may vary in accuracy and quality from clinician to clinician, from ECG system to ECG system, and from site to site, thus making the assessment of multicenter trial results very suspect and prone to false-negative and false-positive responses (Morganroth 2001).

Errors in ECG interpretation are not limited to generalist clinicians or primary care providers. A survey of ECG interpretations conducted in 1993, in London, England teaching hospitals, including doctors in anesthesia, cardiology, and cardiothoracic specialties, revealed widespread errors in correct interval identification of a schematic ECG trace (Montgomery et al. 1994). Of all doctors completing a questionnaire, 74% did not have sufficient knowledge to measure a PR interval, and 76% defined the QT interval incorrectly. The risk of errors in manual measurement may therefore be significant in the absence of retraining and quality control of site estimates during clinical trials.

The use of automated analyses provided by ECG equipment is sufficient to guide the investigator in the immediate assessment of the ECG for major safety findings, but is not often accurate to define the QTc duration and almost universally depicts only the inaccurate corrected QT by Bazett (see below in this section) (Bazett 1920). Inaccuracies are due to the nature of the algorithm and presence of T and U wave abnormalities. Ever since the 1997 European ECG Guidance (CPMP 1997), all ECG intervals require manual measurements in a standardized acceptable manual method by trained personnel.

One of the most troubling problems for clinical research is that of all the ECG intervals, the QT interval is subject to an important large degree of spontaneous variability in its duration, which averages roughly 75 ms in a day (range, approximately 15–150 ms) (Morganroth et al. 1991). Normal values are typically considered to be up to 440 ms, with a range from 365 to 440 ms. The degree of spontaneous change may be enhanced in patients with underlying heart disease and the effect in subjects with oncologic disease and other conditions has not been studied. To reduce the spontaneous variability, attention must be given to basic clinical trial design issues such as sample size, frequency of measurements taken (number of ECGs at baseline and on therapy), accuracy of the ECG interval durations (central validated consistent manual determinations), homogeneity of the study population (healthy volunteers with half women), controls for environmental stresses (activity, food, diurnal effects, time effects, etc.), and especially the effect of heart rate on QT duration.

Another import issue is that the QT interval varies inversely with the magnitude of the heart rate (as the heart rate slows, the QT increases). Thus, it is critical to correct the QT duration for heart rate providing as the ECG interval of interest the corrected QT interval or QTc. The Bazett correction formula (QTcB = QT/square root of the RR interval in seconds), used almost exclusively in clinical practice, overcorrects at elevated heart rates and undercorrects at rates below 60.

Since 1999, in drug development, the Fridericia (Fridericia 1920) formula (QTcF = QT/cube root of the RR interval in seconds) has been used widely since this correction tends to be more resistant to the effects of heart rate changes and thus more accurate. ECG analyses in individuals will generally provide these two fixed exponent QTc durations, and for individual clinical trials data analysis using QTcF is preferred. Individual patient-defined correction formulae (Malik et al. 2002) are by design the most accurate in correcting for the heart rate effect on the QT duration. This is done by taking each individual in a trial and calculating that individual's exponent, which eliminates the influence of heart rate on QT duration. This requires at least 35–50 or more ECGs on placebo or at least at baseline in which these ECGs encompass a range of spontaneous heart rate changes to have enough power to accomplish this task. The individual QTc has been routinely employed in the Thorough ECG Trial and for cytotoxic oncology agents without control groups; with limited sample sizes, the use of this method would not be warranted from a cost–benefit analysis.

13.3 Recent Regulatory Guidances on the Use of ECGs in Clinical Research

With increasing concern about regulatory actions from drugs affecting cardiac repolarization, the United States Food and Drug Administration (FDA) teamed up with Health Canada (2002) to address a more detailed set of recommendations for the drug development industry to consider. The hallmark of this approach was the implied need that a thorough or definitive assessment of a new drug's effect on cardiac repolarization was needed before marketing approval would be granted. This "definitive later changed to thorough" trial to define ECG effects of a new drug first appeared in the November 2002 FDA-Health Canada ECG concept paper entitled "The clinical evaluation of QT/QTc interval prolongation and proarrhythmic potential for non-antiarrhythmic drugs" (FDA and Health Canada 2002). It is important to note that because of the lack of observing TdP or other manifestations such as syncope in a typical 5,000-patient marketing application such as the FDA's New Drug Application (NDA), it is not reassuring that a drug has no cardiac liability, since the 95%

confidence interval for TdP would range from 0 to 1 in 1,600 and 1/1,600 is a very high occurrence rate of this life-threatening adverse event when millions of patients may be exposed in the market. Postmarketing surveillance can help in detecting TdP, but underreporting and ascribing TdP to underlying risks such as heart disease limits this approach (Roden 2004). Thus, the QTc interval duration in clinical development for most drugs had to be assessed in a Thorough or Dedicated ECG Trial to rule out a 5-ms effect. The exception of course would be drug programs that could not use placebo controls and large sample sizes of healthy volunteers, the best example of which is the cytotoxic or genotoxic oncologic products.

Other important provisions of the original FDA Concept Paper to enhance ECG data included:

- ECGs should be recorded, processed, and stored in a digital manner using a centralized ECG laboratory
- Analysis should be done using manual measurement methods though the use of site-automated interval measurements can be used to screen for immediate safety issues

The FDA Concept Paper has now been subjected to the ICH process and a final document should be expected within a couple of years.

Even though the oncologic products that are cytotoxic cannot be used in a Thorough ECG Trial, I will review the basic outline of such a trial to emphasize the standards needed to define the effects on the ECG of most new drugs. In this context, we can then review recommendations for ECG assessments in oncology.

13.3.1 Principles of the Thorough ECG Trial

To reduce the spontaneous variability attention must be given to basic clinical trial design issues such as sample size (usually at least 40 subjects per treatment group), frequency of measurements taken (number of ECGs at baseline, often at least 35–50, and the same number at the same time-points on therapy), accuracy of the ECG interval durations (central validated consistent manual determinations), homogeneity of the study population (healthy volunteers with half women), controls for environmental stresses (activity, food, diurnal effects, time effects, etc.)

and the use of the individually defined QTc duration to eliminate the changes in heart rate from baseline to therapy (Morganroth 2004).

From a clinical point of view, it seems more meaningful to conduct the Dedicated ECG Trial in the target population for the drug's use; however, selecting patients with clinical diseases will make the ECG results subject to patients' multiple degrees of disease intensity, multiple comorbidities, and different concomitant medications. Balancing all these factors, which can effect the duration of the QTc interval, would require too large a sample size and too difficult a recruitment to make the ECG trial definitive. Since the effect of drugs on cardiac intervals will occur in healthy as well as abnormal hearts or other clinical conditions, it is more effective to select a homogeneous disease and drug-free group of healthy volunteers as the study population. To mimic the new drug's interaction with any effect modifiers that might be present in the target population (e.g., heart disease, metabolic abnormalities, concomitant metabolic inhibitors or abnormal metabolism) a supratherapeutic dose in the healthy volunteers must be employed, which will also provide a dose–response relationship. Usually, the minimal clinical dose compared to the supratherapeutic dose is at least three to five times apart and for certain agents such as antihistamines or antibiotics tends to be over ten times apart.

Important considerations in characterizing the dose–response or concentration–response relationship include:

- The maximal degree of the QTc prolongation
- The steepness of the slope between QTc prolongation and dose/concentration
- The relationship between the threshold dose for QTc prolongation and the therapeutic dose range, linearity or nonlinearity of the dose–concentration effect dependency, and the time course of QTc prolongation in relation to plasma levels

Furthermore, to interpret the results of a thorough ECG trial, adequate control groups are required. Without a placebo, it is difficult to determine the effects of spontaneous variability. Reasons for inadequate control may be too few ECGs, too few subjects, poor quality in ECG duration measurements, poor control of activity and food, bad timing for ECG time points, etc.

The most significant new recommendation by the FDA, which has been continued unchanged in the ICH process for the conduct of a Thorough ECG Trial, is the required inclusion as a control group of a "positive control." The positive control group should demonstrate a small 5–10 ms change from baseline in the QTc duration, which is similar to the magnitude of the threshold of defining a new drug as having a QTc effect. This magnitude is considered by the regulators to be the threshold where risk of TdP begins and thus if one claims the trial shows no difference between the new agent and placebo it is critical to be certain the trial had sensitivity to rule out a 5-ms effect by showing that a positive control arm demonstrated that effect in the same trial under the same conditions.

13.4 Clinical Problems with Assessing ECG Effects of Cytotoxic Oncology Drugs

As noted from above, the difficulties in assessing the ECG effects of cytotoxic oncologic agents are readily apparent. The lack of studying healthy volunteers under a variety of doses with negative and positive controls with adequate sample sizes limits the assessment of ECG changes. Thus, in the usual trail, especially in phase I, all changes seen on the ECG will be naturally attributed to the new drug. Even if another co-morbid condition is present that may cause the observed change, the issue of whether the new drug is a contributor to that effect will be raised. Since there are many drugs and co-morbid conditions that can cause an effect on cardiac repolarization, trying to minimize these confounding conditions helps to be more specific in terms of the new drug's ECG effects.

QTc durations can increase with:

- Electrolyte abnormalities, especially hypokalemia and hypomagnesemia
- Bradycardic states
- Myocardial ischemia, infarction, heart failure, and hypoxia
- Central nervous system disorders
- Endocrinopathies, especially hypothyroidism
- Congenital long QT syndrome

Drug-induced QT prolongation covers a wide array of therapeutic classes including antiarrhythmic drugs (especially amiodarone and sotalol), psychotropic agents, antibiotics, etc.

Even using positive control agents in phase II and III oncology trials may not help, since many other oncologic agents have not been fully characterized as to their ECG effects.

Thus the ECG analysis in an oncologic trial will rely primarily upon the observed effects in the study population and thus adequate attention to the following principles as defined in Sect. 13.2 need to be addressed:

- Sample size: make it as large as possible
- ECG analysis: use a central validated ECG laboratory
- Let QTcF and not QTcB be the final analysis even if the site makes decisions on QTcB (try to have central ECG QTcF determinations drive site decisions when possible)
- Minimize confounding factors as noted
- Adequate determination of ECG data requires adequate frequency of ECG assessments.

13.4.1 Adequate ECG Frequency in Clinical Oncology Trials: Based on Statistical Issues

For ECG data analysis, the primary method of defining the ECG results is to calculate the mean change from baseline for all subjects in the same dose regimen. This produces the treatment effect. Of course the mean change "hides" outliers, and therefore a categorical or outlier analysis is critical. One form is the use of the maximum mean change in all treatments, which produces comparable data in my experience. This is done by looking at the largest positive change from baseline on treatment at any time point (average the ECGs around that time point for better time point precision) in each subject and then calculate the mean maximum change for all subjects in each treatment arm.

The outlier or categorical analysis is defined as determining what percentage of the patients on each treatment demonstrate a change from baseline in QTc (and all other ECG intervals) that is of a particular magnitude that identify them as being at potential risk because of the QTc effect. This is done by taking the largest positive change form the baseline value for each ECG interval at whatever time point on treatment that this occurs (take the mean of all the ECGs used around that time point to establish that point's ECG interval value). Most clinicians use a QTc of 500 ms or more in duration as a clinical risk requiring reevaluation of drug dose or risk benefit of the therapy (Priori 2003). Thus, the specific clinical criterion is a new greater than 500-ms QTc duration or the observation on the drug or an abnormal T-U wave, often thought to represent an early afterdepolarization that may be a harbinger of TdP. Statistically, a change in QTc of more than 60 ms from baseline in an individual is considered as a specific outlier criterion. An often overly sensitive (too many subjects on placebo will show this effect) criterion is a 30- to 60-ms change from baseline (Pratt et al. 1996; Morganroth et al. 1993). The use of other QTc outlier criteria such as normal to abnormal, percentage of subjects over 480 ms, etc., in my opinion adds little except that the more times you look the more the likelihood of a false-positive response. I believe for QTc the criteria to look at are:

- 30–60, > 60 ms change from baseline
- New absolute 500 ms
- New abnormal U waves

Of course all the ECG data are expressed in terms of means, standard deviations, ranges, and confidence intervals.

Since all ECG data are in the form of a change from baseline, the measurement of the baseline ECG intervals critically influences the results. Even for routine phase I–III clinical trials in general, the FDA–Health Canada Concept Paper notes that the "use of baseline values from single ECGs is a practice to be discouraged." What is recommended is that baseline ECGs should be computed as the mean of multiple ECGs to enhance the precision of the measurement in light of the large degree of spontaneous variability. It is my opinion that in routine clinical trials, at least three baseline ECGs should be obtained, and in cytotoxic oncology trials, perhaps six ECGs should be used to define the best point estimate of the baseline ECG interval values, especially for QTc. These baseline ECGs can be taken, even if only a few minutes apart, to reduce the QTc variance.

It is critical that the ECG sampling frequency on treatment covers the extent of exposure of the drug and its metabolites and also accounts for late effects that may occur for some drugs. Thus in cycle 1 and probably in cycle 2 and possibly in cycle 3, an intense frequency of on-therapy ECGs must be obtained. Intense is defined as perhaps 2, 4, 6, 8, 12, and 24 h after dosing or especially when the pK samples are drawn so that a full pK-pD relationship can be defined. For late effects, perhaps 2 additional days of ECG data should be sufficient. In later cycles and in phase III, the frequency of ECGs obtained should follow the principle that whenever you obtain a measurement of an organ's status (e.g., bone marrow or liver by blood tests), then an ECG to check on cardiac safety seems reasonable (especially since the ECG is noninvasive and costs less).

13.5 A New Technology to Record Standard 12-Lead ECGs

In addition to standard resting supine ECGs, new Holter technology allows for continuous ambulatory recording of 12-lead digital ECGs for 24 h at a time, thus making it easy in some settings to obtain the frequent ECG data in phase I trials. This Holter approach should prove useful for the assessment of cardiac repolarization at numerous discrete time points following drug administration or at baseline. It also has additional applications for continuous beat-to-beat analysis, as well as for detecting cardiac arrhythmias (traditional Holter monitoring). Validation that the 12-lead ECGs recorded by a Holter device rather than a standard ECG recorder is required to be certain that the new technology is comparable to the historic standard. Such as study has been conducted as well as validation in accordance with *21CFR11*. This study compared the utility for QTc risk assessment of ECGs recorded by standard or digital 12-lead Holter devices, as well as the precision of QT and RR interval measurement by the manual digitized systems (digitizing board and digital on-screen calipers) on standard and Holter-derived ECGs (Sarapa et al. 2004). The results of the study showed that the QT, QTcF, and RR data produced by manual measurements on standard ECGs were essentially equivalent to those from digital 12-lead Holter ECGs.

13.6 What Is the Regulatory Implication of Finding a Drug Prolongs Cardiac Repolarization?

For most drugs, there is regulatory guidance on how great a magnitude of a QTc duration effect might be related to risk of TdP. This analysis has come from the experience of regulators over the past decade, mostly from approving QTc prolonging drugs and observing their effect in the market. There is general consensus that a magnitude of QTc duration effect of for nononcologic products is (Shah 2002):

- 0–5 ms, imparts no risk of TdP.
- If the effect is over 20 ms, the risk is considered quite high for TdP.
- 5–10 ms effect for a drug is of minimal concern but depends on the risk–benefit ratio of the particular drug. An effect between 10 and 20 ms is uncertain.

The only published experience to date for an oncologic product with a demonstrated QTc effect is the drug arsenic trioxide (Barbey et al. 2003). That agent had its efficacy established before developing an assessment of the ECG effects of the agent. In fact, the early assessments did not show any major ECG effects. The report cited was in a series of 99 patients with 170 courses of therapy retrospectively collected for a central ECG analysis. The presence of a baseline (one ECG) was found in only 56 of the 99 subjects and there were complications of metabolic abnormalities and concomitant QT prolonging medications. Nevertheless, this agent showed that 68% of the subjects had a prolongation of the QTc and in 46% the change was greater than 500 ms; 35% had a greater than 60-ms change from baseline. The central tendency effect was around 47 ms. Because of this agent's benefit and the ability to advise physicians how to monitor for ECG changes on the label, this drug was approved albeit with a black-box warning in reference to the QTc effects.

The regulatory impact of the QTc changes observed on a cytotoxic oncology product will be based on a risk–benefit decision after a complete evaluation of the ECG effects during the drug's development. Attention will be given to the PK–PD relationship to see the magnitude of the QTc change and its slope (flatter the slope the less worrisome the response) as well as the proportion of outliers, especially those over a 60-ms change from baseline and new absolute results over 500 ms. Clinical cardiac findings such as syncope, ventricular tachyarrhythmias (especially TdP), and other cardiac effects will be important in this analysis. In the end, if the drug prolongs life, then a cardiac ECG change may not necessarily be grounds for disapproval as long as the characteristics of the ECG change are well described and can be used to construct a set of labeling instructions to minimize this side effect's potential impact.

References

Barbey JT, Pezzullo JC, Soignet SL (2003) Effect of arsenic trioxide on QT interval in patients with advanced malignancies. J Clin Oncol 21:3609–3615

Bazett H (1920) An analysis of the time-relation of electrocardiograms. Heart 7:353–370

CPMP (1997) Points to consider: the assessment of the potential for QT interval prolongation by non-cardiovascular medicinal products. EMEA, London

Food and Drug Administration and Health Canada (2002) The clinical evaluation of QT/QTc interval prolongation and proarrhythmic potential for non-antiarrhythmic drugs. http//www.fda.gov/cder/workshop.htm#upcoming

Fridericia LS (1920) Die systolendauer im elektrokardiogramm bei normalen menschen und bei herzkranken. Acta Med Scand 53:469–486

Malik M, Färbom P, Batchvarov V et al (2002) Relation between QT and RR intervals is highly individual among healthy subjects: implications for heart rate correction of the QT interval. Heart 87:220–228

Montgomery H, Hunter S, Morris S et al (1994) Interpretation of electrocardiograms by doctors. BMJ 309:1551–1552

Morganroth J, Brozovich FV, McDonald JT, Jacobs RA (1991) Variability of the QT measurement in healthy men: with implications for selection of an abnormal QT value to predict drug toxicity and proarrhythmia. Am J Cardiol 67:774–776

Morganroth J, Brown AM, Critz S, Crumb WJ, Kunze DL, Lacerda AE, Lopez H (1993) Variability of the QTc interval: impact on defining drug effect and low-frequency cardiac event. Am J Cardiol 72:26B–32B

Morganroth J (2001) Focus on issues in measuring and interpreting changes in the QTc interval duration. Eur Heart J Supplements 3 [Suppl K]:K105–K111

Morganroth J (2004) Design and conduct of the Thorough Phase I ECG trial for new bioactive drugs. In: Morganroth J, Gussak I (eds) Cardiac safety of noncardiac drugs. Humana Press, Teterboro, NJ, pp 205–223

Priori SG, Schwartz PJ, Napoliatano C et al (2003) Risk stratification in the long QT syndrome. N Engl J Med 348:1866–1874

Pratt CM, Ruberg S, Morganroth J et al (1996) Dose-response relation between terfenadine (Seldane) and the QTc interval on the scalar electrocardiogram: distinguishing a drug effect from spontaneous variability. Am Heart J 131:472–480

Roden D (2004) Drug-induced prolongation of the QT interval. N Engl J Med 350:1013–1022

Sarapa N, Morganroth J, Couderc J-P et al (2004) Drug-induced QT prolongation in the electrocardiogram: assessment by different recording and measurement methods. Ann Noninvasive Electrocardiol 9:48–57

Shah RR (2002) The significance of QT interval in drug development. Br J Clin Pharmacol 54:188–202

Yeh E, Tong A, Lenihan D et al (2004) Cardiovascular complications of cancer therapy. Circulation 109:3122–3131

14 The Role of PET Scanning in Determining Pharmacoselective Doses in Oncology Drug Development

P. Price

Abstract. Molecular imaging is the most sensitive and specific method for measuring in vivo molecular pathways in man. Its use in oncology has developed significantly over the last 5–10 years. Molecules can be labelled with positron emitting isotopes and the emitted radiation is detected using sensitive positron emission tomography (PET) cameras. It is now possible to measure in vivo and normal tissue pharmacokinetics of anti-cancer drugs and investigate their mechanism of action. Radiolabelling of tracers can be used to measure specific pharmacodynamic endpoints and target identification. Increasing evidence shows how these technologies, when added to early drug development, can rapidly reduce the time for entry into man and early identification of mechanisms of action. With the move towards more segmented markets and identification of specific subgroups, PET's use for noninvasive biomarkers will become increasingly important. However, much international effort between academia and industry is required with prioritisation of development of this technology.

14.1 Background to Molecular Imaging

PET is one of the most sensitive and specific methods for measuring in vivo processes in humans (Jones 1996). It is a technology that can link molecular, cellular and animal studies into studies of diseases of tissues in humans. Molecules are labelled with positron emitting isotopes produced from the cyclotron, and following injection into patients, the activity is detected in patients. Rate kinetics can be obtained providing information on the movement of molecules within the living tissue (Price 2001).

14.2 Preclinical Work

Much funding has been devoted to the development of small animal imaging over the last 5 years. The advances have been in developing more sensitive small-animal imaging cameras with down to 3 mm resolution. Other areas of molecular imaging have also developed preclinically, such as bioluminescence, MRI, etc. While these types of studies can be used in their own right for recording biological information, they can also be used to develop paradigms prior to clinical study in man. The main advantage is that animals can be imaged repeatedly and noninvasively on the scanner, which will decrease the number of animals required for experiments and increase the statistics of our experiments. There are also advantages for taking measurements in deep-seated tumours. Many companies now are investing in their own systems to compliment their preclinical drug development.

14.3 In Vivo Pharmacokinetics

Radiolabelling of anti-cancer drugs is now possible and recent developments in radiochemistry mean that many anti-cancer molecules can now be labelled with positron emitting isotopes. Work performed in collaboration with Cancer Research UK involved the radiolabelling of the drug temozolomide. The pharmacokinetics of the molecules in brain tumour patients has been assessed and pharmacokinetic parameters have been obtained (Mieckle et al. 1998) (Fig. 1, Table 1). Information on the relationship between in vivo pharmacokinetics and response can now be obtained. This type of information will become increasingly important with targeted therapies where we cannot rely on the plasma concentration to provide tumour and normal tissue information.

Work has also been performed looking at the mechanism of action of compounds. One experiment conducted with Cancer Research UK and Glaxo Wellcome involved the use of the 5FU metabolism inactivator eniluracil and its effects on the in vivo tumour and normal tissue

Table 1. Pharmacokinetics of temozolomide

	Normal grey matter	Tumour
VD_{part}	6.97	54.40
K_1 (min^{-1})	0.18	1.44
MRT (min)	35.68	37.27
$T_{1/2z}$ (min)	26.63	32.20

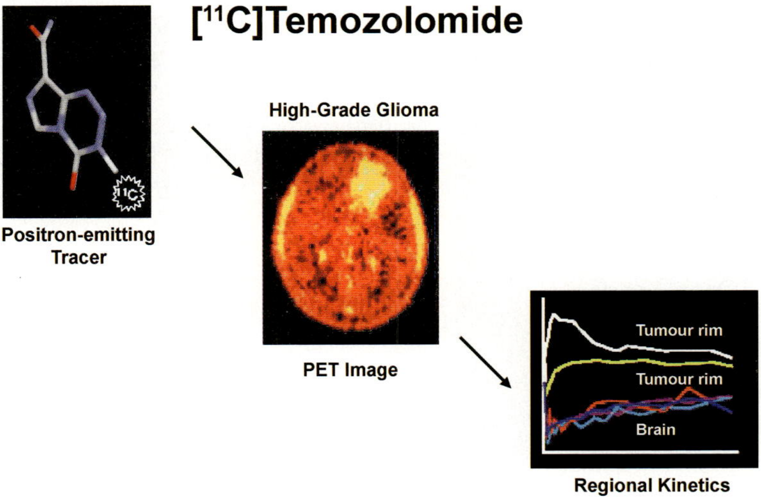

Fig. 1. Radiolabelled temozolomide

pharmacokinetics of ^{18}F-labelled 5FU (Saleem et al. 2000). Figure 2 demonstrates the difference in distribution of the radiolabel and the increase in uptake of 5FU following DPD activation in tumours.

PET pharmacokinetics of anti-cancer drugs can also be used in the pre-phase I setting: so-called microdosing. Legislation for the fast entry of compounds into humans is being developed under the FDA. A PET study conducted between Cancer Research UK and the MRC Cyclotron Unit showed how a new compound can be entered into humans prior to

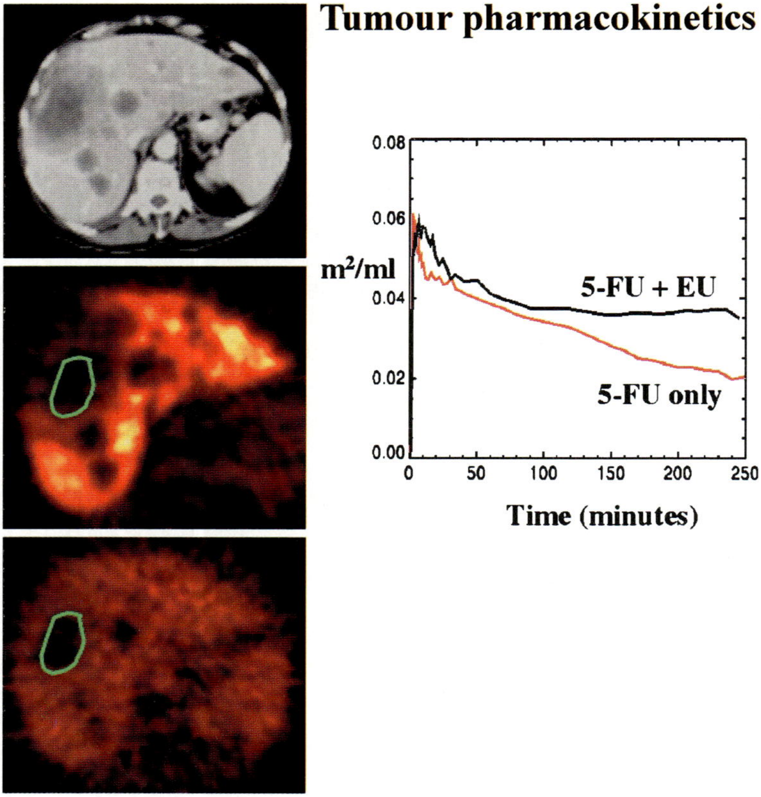

Fig. 2. Tumour pharmacokinetics of 5-fluorouracil (5FU) vs 5-fluorouracil plus ethynyluracil (5FU + EU)

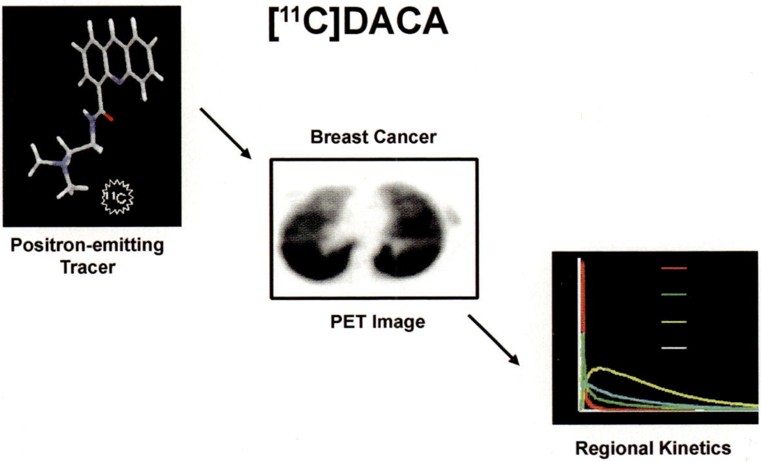

Fig. 3. Radiolabelled DACA

phase I and the type of information that can be obtained (Fig. 3). One year prior to phase I, the radiolabelled molecule was injected in patients as a 1/1,000th of the phase I starting dose, and information about the plasma tissue and tumour pharmacokinetics was obtained (Saleem et al. 2001). This experiment could be repeated with analogues of the compound and structure–function relationships investigated at microdosing levels of compound, which should accelerate drug development significantly, as in vivo structure–function relationships can be investigated in patients (Osman et al. 2001).

14.4 Pharmacodynamics

There are a number of approaches that can be used for investigation of in vivo pharmacodynamic endpoints in patients. ^{18}F-FDG is an analogue of glucose which is now used routinely in staging patients. Figure 4 describes the mechanism by which it can be used to detect tumours. FDG is taken up in Glut1 receptors which are overexpressed on tumour cells and trapped within the cell by metabolism by hexokinase. This creates a high signal in the tumour compared to normal tissue. This is

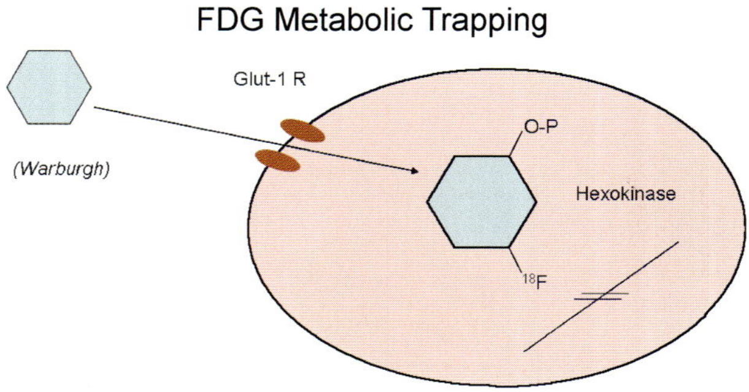

Fig. 4. FDG metabolic trapping

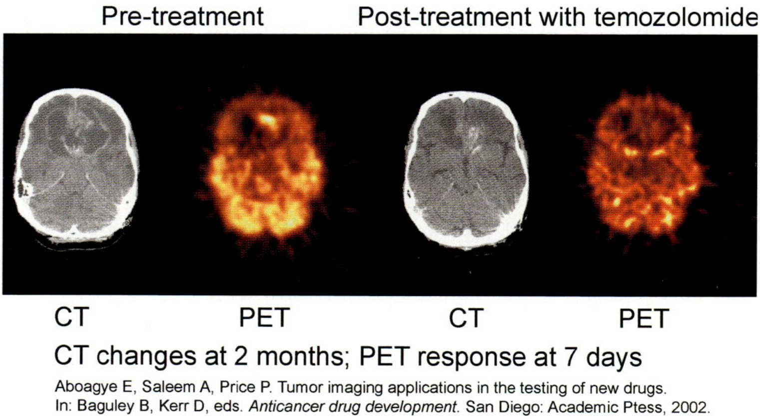

Fig. 5. CT and PET brain tumour scan images before and after treatment with temozolomide

the basis of the technique that has been used to develop FDG PET for staging in cancer. However, quantitative measurements can also be taken and used to assess patient response. Figure 5 demonstrates the response assessment in patients with brain tumours in response to temozolomide. A radiological response can be detected at 2 months, whereas an FDG response can be measured at 7 days (Brock et al. 2000). PET is being increasingly used for quantitative assessment response to therapy and

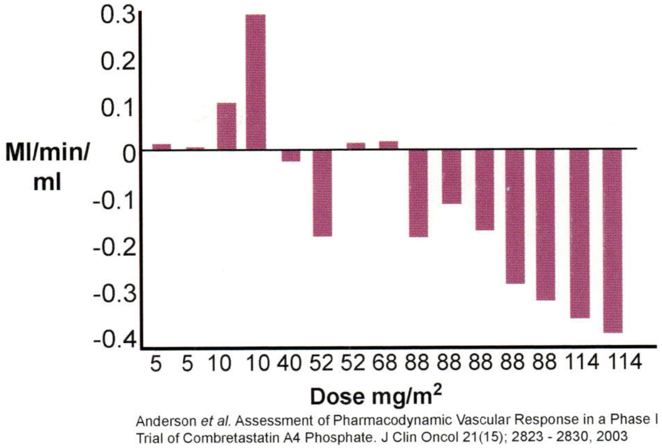

Fig. 6. Absolute change in tumour perfusion

guidelines have been produced to recommend its uses in this situation (Young et al. 1999). More novel compounds can now be used to assess other pharmacodynamic responses such as ^{11}C thymidine (Wells et al. 2003) and $H_2^{15}O$ to assess flow response (Anderson et al. 2003).

A phase I trial of an antivascular agent combretastatin has been conducted with parallel PET studies investigating the four physiological parameters of blood flow, blood volume, volume of distribution and cardiac output. The parallel PET studies provided this pharmacodynamic information within a phase I setting (Anderson et al. 2003) (Fig. 6). Such work involves the assessment of the mechanism of action of the compound in phase I. An accompanying editorial to this publication by Jerry Collins of the FDA(Collins 2003) discussed how functional imaging in phase I studies could help decision making.

14.5 Conclusion

There have been a number of lessons learned on the use of molecular imaging to assist in early clinical trial development. Certainly, advice should be sought early from professional groups, and planning early in the drug development phase is vital.

Major challenges in the design of such studies include confirmation of presence of the target, confirmation of target hit, quantitation of effects of the drug-target interaction and assessment of the optimal biological dose. In vivo pharmacokinetics can be used for in vivo drug selection and to study how this can be optimised.

To move this field forward, the National Cancer Institute have undertaken a number of workshops that are summarised at http://cip.cancer.gov (National Cancer Institute 2005).

An international effort between academia and industry is required to move this field forward with prioritisation of clinical questions and development of methodology. An institute for such studies has recently been established in Manchester with support from Cancer Research UK and hopefully will contribute to the international effort of developing this technology to assist in our rational drug development, speed up the process, and therefore save much time and money.

References

Anderson HL, Yap JT, Miller MP, Robbins A, Jones T, Price PM (2003) Assessment of pharmacodynamic vascular response in a phase I trial of combretastatin A4 phosphate. J Clin Oncol 21:2823–2830

Brock CS, Young H, O'Reilly SM et al (2000) Early evaluation of tumour metabolic response using [^{18}F]fluorodeoxyglucose and positron emission tomography: a pilot study following the phase II chemotherapy schedule for temozolomide in recurrent high-grade gliomas. Br J Cancer 82:608–615

Collins JM (2003) Functional imaging in phase I studies: decorations or decision making? J Clin Oncol 21:2807–2809

Jones T (1996) The role of positron emission tomography within the spectrum of medical imaging. Eur J Nucl Med 23:207–211

Meikle SR, Matthews JC, Brock CS et al (1998) Pharmacokinetic assessment of novel anti-cancer drugs using spectral analysis and positron emission tomography: a feasibility study. Cancer Chemother Pharmacol 42:183–193

National Cancer Institute (2005) Cancer Imaging Program http://cip.cancer.gov

Osman S, Rowlinson-Busza G, Luthra SK et al (2001) Comparative biodistribution and metabolism of carbon-^{11}-labeled N-[2-(dimethylamino)ethyl]acridine-4-carboxamide and DNA-intercalating analogues. Cancer Res 61:2935–2944

Price P (2001) PET as a potential tool for imaging molecular mechanisms of oncology in man. Trends Mol Med 7:442–446

Saleem A, Yap J, Osman S et al (2000) Modulation of fluorouracil tissue pharmacokinetics by eniluracil: in-vivo imaging of drug action. Lancet 355(9221):2125–2131

Saleem A, Harte RJ, Matthews JC et al (2001) Pharmacokinetic evaluation of N-[2-(dimethylamino)ethyl]acridine-4-carboxamide in patients by positron emission tomography. J Clin Oncol 19:1421–1429

Wells P, Aboagye E, Gunn RN et al (2003) 2-[11C]thymidine positron emission tomography as an indicator of thymidylate synthase inhibition in patients treated with AG337. J Natl Cancer Inst 95:675–682

Young H, Baum R, Cremerius U et al (1999) Measurement of clinical and subclinical tumour response using [^{18}F]-fluorodeoxyglucose and positron emission tomography: review and 1999 EORTC recommendations. European Organization for Research and Treatment of Cancer (EORTC) PET Study Group. Eur J Cancer 35:1773–1782

15 Biometrical Aspects of Drug Development

D. Machin, S-B. Tan

Abstract. Once the activity of a compound has been established in the laboratory (usually by use of experimental animals) the next stage of development is to bring this forward to humans in early-phase clinical trials. A pharmacokinetic study aims to establish an effective dosing regimen for the compound in order to reach concentrations within the therapeutic window as quickly as possible. The aim of the phase I trials is typically to determine a maximal safe dose with which more rigorous investigation of activity in a phase II trial can be conducted. This chapter deals with statistical issues related to the design of phase I studies.

15.1 Introduction

The design of any study is the key component for obtaining a satisfactory answer to the question posed. An important factor when considering the design of phase III trials is the information provided from earlier-stage studies and trials. Consequently, since the development of, for example, new therapies tends to progress at the clinical stage through pharmacokinetic, phase I to phase II then to phase III trials, the sequential nature of this structure implies that reliable information from one step is important for the next. Poor experimentation at the relevant stage can clearly jeopardise the design of the next stage and, at best, results in a waste of resources and at worst may compromise patient safety. Unfortunately, the evidence provided by published reports of early-stage trials suggests that these have often not been well designed or well reported.

A phase I trial aims to determine (often from a preselected range of potential doses) the dose that can be utilised at the next stage of development and so focuses on selecting the highest practical dose, the presumption being that the greater the dose the greater the anti-disease effect will be. However, safety considerations dictate that the dose chosen for the subsequent trials should have an acceptable toxicity profile.

Phase I trials are usually small. This lack of numbers has a least two implications, one is that the final estimates of whatever statistic is to be estimated will be rather imprecise and the other is that these studies should be very carefully designed so as to maximise the information that can be obtained by optimal use of this scarce resource.

15.2 Phase I Trials

In broad terms, the aim of a phase I trial is to establish the maximum tolerated dose (MTD) of a particular compound. In some circumstances, the treatment under test may prove to be too toxic and so no MTD is established. In this case, a phase II trial would not be initiated for subsequent further testing. Underestimation of the MTD may lead to an apparent lack of efficacy at the later stages. Overestimation may lead to unacceptable toxicity (even death) in some patients. In either situation, a potentially useful compound may be shelved and opportunities for a therapeutic advance stalled.

For patients with a specific disease, one objective of treatment may be to reduce (or eradicate) the disease burden. However, it is recognised that any attack on the disease itself by a chemotherapeutic or other agent may bring collateral damage to normal tissue and vital organs. The usual strategy is to attempt to balance the two by establishing the concept of dose limiting toxicity (DLT) by means of a phase I trial assuming an underlying theoretical dose response curve. Once the dose at which DLT occurs is established, the MTD is then defined as one (dose) step down from this dose

The level and type of DLT may be very specific to the clinical situation under investigation but should be defined before the phase I trial commences. Once so determined, the presence or absence of such toxicity is recorded carefully when a patient receives the compound under study. However, this presumes that a first dose (say d_{START}) has been identified for the design and that this has been given to the first patient.

Thus in advance of the first patient being recruited in a phase I trial, the investigators first identify the range of doses to use and all the specific-dose levels to test. Thus d_{START} will be one of these options and the ultimately identified MTD will also be one of these predefined doses. There are some difficulties with such an approach, since one is likely to start at low dose and then proceed dose-step by dose-step to successively higher doses.

Initially the minimum dose to investigate, d_{minimum}, is determined, then attention naturally turns to establishing what might be considered the therapeutic range and the setting of the maximum dose, d_{maximum}, for the study. Once these are established, then the intermediate doses to investigate can be established.

For convenience, we label the k doses finally chosen as $d_1 = d_{\mathrm{minimum}}$, $d_2, d_3, \ldots, d_k = d_{\mathrm{maximum}}$. However, we still need to choose k and the specific values for each of the intermediate doses between the minimum and maximum values already defined. Statistical design considerations may suggest that these should be chosen equally spaced between d_{minimum} and d_{maximum} on either a linear or a logarithmic scale. The doses may depend on how the drug is packaged – perhaps in tablet form or vial of a certain volume where dose choice may be limited or in a powder or liquid more easily constituted into any dose.

However, practice has often recognised that as the dose increases in equal steps it may become sequentially more and more toxic and hence possibly dangerous for the well-being of the patient. This caution has then led many investigators to decrease the step sizes as the dose increases. One method uses the Fibonnacci series. Fibonnacci (c. 1180–1250) was an Italian mathematician who first studied the following mathematical series: $a_0 = a_1 = 1$, then from a_2 onwards $a_{n+1} = a_n + a_{n-1}$. This gives the series: 1, 1, 2, 3, 5, 8, 13, 21, 34, etc. The corresponding Fibonacci ratios of successive terms are: $1/1 = 1$, $2/1 = 2$, $3/2 = 1.5$, $5/3 = 1.667$, $8/5 = 1.600$, $13/8 = 1.625$, $21/13 = 1.615$, $34/21 = 1.619$, ..., and eventually as n gets larger and larger this approaches $1.33 = 2/(\sqrt{5} - 1)$. These ratios are shown in Table 1 and, for relatively small n appropriate to the number of dose levels in a phase I study, the ratio oscillates up and down. In mathematical terminology, the series of ratios is not monotonically decreasing and so in fact do not provide successively decreasing step sizes. There is no theoretical reason why this or any other mathematical series should be chosen for this purpose – they are merely empirical devices.

Nevertheless, it is usually regarded as desirable that successive doses are a decreasing multiplier of the preceding dose and thus (often without a clear explanation provided) *modified* Fibonnnaci multipliers like those in Table 1 are substituted in practice. However, it is usually pragmatic considerations that determine the modifications and no systematic rationale underlies the changes.

Table 1. Dose-escalation methods based on the Fibonnacci series and that used for a phase I study of nolatrexed dihydrochloride conducted by Estlin et al. (2001)

	Fibonnacci ratio		Nolatrexed dihydrochloride	
Dose	Full	Modified	Escalation	Dose ($mg/m^2/day$)
d_1	1	1	1	600
d_2	2	2	**1.33**	800
d_3	1.50	1.67	1.20	960
d_4	1.67	1.50	1.17	1,120
d_5	1.60	1.40	1.07	1,200
d_6	1.63	1.33	1.20	1,440
d_7	1.62	1.33	1.11	1,600
...	...	...		
d_∞	**1.33**	**1.33**		

15.2.1 C33D

A common, sequential design is to choose a low starting dose, perhaps with $d_{\text{START}} = d_{\text{minimum}}$ and a fixed number of replicates (often 3). The choice of the next dose, d_{NEXT}, then depends on the number of patients (0, 1, 2 or 3) experiencing DLT. Clearly, if no patients experience DLT, then the subsequent dose to investigate will be higher than that just tested. This process continues until either the stopping level of DLT is attained in the successive groups of three patients or d_{maximum} has been tested. In circumstances where the first two patients both experience DLT at a particular dose, it is not usual to give the third patient this same dose but to change the dose chosen to a lower one from the prespecified dose range. Using this type of strategy, Smith et al. (1998) state that the MTD from a phase I design is established by adding cohorts of three patients at each dose level, and using the rules in Fig. 1 to determine whether dose escalation should occur. This is known as a cumulative "3+3" dose (C33D) method.

Although this process will (in general) establish the MTD, it is only a pragmatic consideration that dictates that the phase I trial should have tested at least six patients at d_{MTD}. This usually implies (as indicated in Fig. 1) that once first identified, extra patients are then recruited and tested at this provisional d_{MTD} until six patients in total have experienced this dose.

Storer (2001) describes a modification to the C33D design by adding a stage before that design is implemented. The strategy is essentially to start the process at a more informative dose than d_{minimum}. This ad-

Commencing with the lowest dose:

(a) if zero of three experience DLT, then escalate to the next higher dose level.

(b) if one of three patients experience DLT, then add three more patients at that dose level:

(i) if zero of these three patients experience DLT (ie only one of six patients at the dose level), then escalate to the next higher dose level.

(ii) if one (or more) of these three patients experience DLT, then the MTD has been exceeded; three more are then added at the previous dose level (if only three patients had been treated previously at the prior level).

Fig. 1. Establishing the MTD in a C33D for a phase I trial (after Smith et al. 1998) junct design suggests recruiting single individuals (rather than three) to successive doses and moving up and down the dose escalation scale according to whether or not a DLT is observed. The design moves into C33D once the current patient has not experienced a DLT *and* one previous patient has experienced DLT *and* one has not.

The C33D design, with or without the Storer (2001) modification, has no real statistical basis, and more efficient alternatives have been sought. Efficiency here can be thought of as achieving the right MTD and with as few patients as possible. However, the design is easy to implement and requires little (statistical) manipulation – only keeping a count of the number of patients experiencing DLT at each dose tested.

15.3 Continual Reassessment Method

O'Quigley et al. (1990) have proposed the continual reassessment method (CRM). This design recruits the first patient to a dose closer to the centre of the range of prespecified doses than the d_{minimum} of C33D. Essentially, if DLT is observed in this first patient then the next patient (patient 2) is given the dose below d_{START}, whereas if no DLT is observed he or she receives the dose above d_{START}. Once this second patient receives the corresponding dose, and presence or absence of DLT is observed, the subsequent dose to utilise (which may be below, at or above the dose last used) is determined. However, at any stage of this process, the results from all individual patients so far recruited are utilised to provide the basis for the choice of the dose to be tested in the next patient recruited.

The same process of selecting the range and actual dose in the C33D design is necessary for the CRM design. In addition, however, to implement CRM it is necessary to attach to each of these doses (based on investigator opinion) the probability of patients experiencing DLT at that dose. We label these probabilities $\theta_1, \theta_2, \theta_3, ..., \theta_k$. This *prior* elicitation of investigator opinion about toxicity leads to CRM being termed a Bayesian design.

It is implicit in the method of selecting these probabilities that, once they are assigned, then a 'reasonable' starting dose, d_{START}, would correspond to the dose that gives a value of the probability θ_{START} close to some 'acceptable' value. This probability is often chosen as less than

0.3. The 0.3 arising as a less than 1 in 3 chance, the '3' coming from history associated with the use of C33D. The chosen d_{START} would not usually correspond to the extremes d_{minimum} or d_{maximum} of the dose range cited.

In the phase I study of Flinn et al. (2000), summarised in Table 2, a dose escalation strategy was utilised with decreasing multiples of the previous dose used. They defined minimum, $d_{\text{minimum}} = 40$, and maximum, $d_{\text{maximum}} = 100$, doses with six $10\,\text{mg/m}^2$ steps. A CRM-based design was used and the investigator's prior probabilities attached to each dose are given in Fig. 2, which (for illustration) is superimposed onto the idealised dose response curve (see below). As might be expected, as the dose is increased the anticipated probability of DLT increases, so that with dose $40\,\text{mg/m}^2$, θ is only 0.05 (or anticipated to be seen in 1 in every 20 patients with this dose), whereas at dose $100\,\text{mg/m}^2$ θ is 0.8 (four in every five patients).

The $d_{\text{start}} = 50\,\text{mg/m}^2$ chosen corresponding to the prior probability of toxicity θ close to 0.1 and a total of 20 patients were eventually included. Their final conclusion was that in patients with advanced non-Hodgkin's lymphoma (NHL), the MTD for liposomal daunorubicin was $70–80\,\text{mg/m}^2$.

The CRM uses a mathematical model for the idealised dose response curve in Fig. 2 and one model (referred to as the Tanh model) for this

Table 2. DLT observed in patients with advanced non-Hodgkin's lymphoma treated with liposomal daunorubicin with constant doses of CVP (after Flinn et al. 2000)

Daunorubicin	Dose escalation	Prior probability of DLT, θ	*z*-score	Number of patients recruited	Number of patients with DLT
40	–	0.05	−1.47	–	–
50 (start)	1.25	0.10	−1.10	4	0
60	1.20	0.20	−0.69	4	1
70	1.17	0.30	−0.42	3	0
80	1.14	0.50	0.00	7	2
90	1.13	0.65	0.31	2	2
100	1.11	0.80	0.69	–	–
		Total		20	5

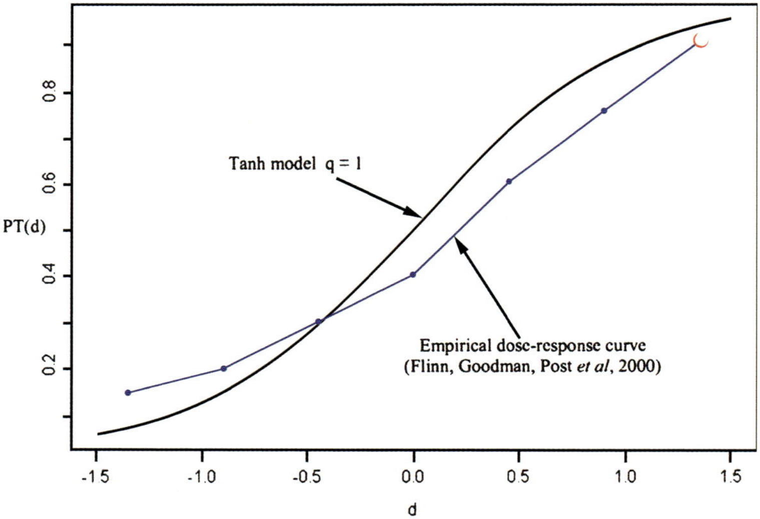

Fig. 2. The first (no data) model for the phase I trial of Flinn et al. (2000)

defines the probability of DLT at dose d as:

$$\mathrm{PT}(z) = \left(\frac{1}{1 + \exp(-2z)}\right)^q \tag{15.1}$$

Here $\mathrm{PT}(z)$ is the probability of DLT at 'working' dose z.

To begin the implementation of the CRM design, the parameter q is set to 1 in (15.1) and the working doses z are selected as

$$z = \frac{1}{2} \log\left(\frac{\theta}{1 - \theta}\right) \tag{15.2}$$

Effectively, this equation converts the proportion θ into the area under a logistic distribution with mean zero and standard deviation 1. For example if $\theta = 0.5$, $z = 0$, $\theta = 0.025$, $z = -1.96$, whereas if $\theta = 0.975$, $z = +1.96$. The corresponding scores for the study of Flinn et al (2000) are given in Table 3.

The investigators then choose a starting dose, a subject is given that dose, and whether or not a DLT occurs is noted (we denote this result as

Table 3. Updated values for z based on ($y_1 = 0$ or 1) at first working dose $z = -0.69$. If $y_1 = 0$ the dose is increased – shaded panels. If $y_1 = 1$ the dose is decreased – open panels

Dose (mg/m^2)	d	40	50	60	70	80	90	100
Probability of DLT	θ	0.05	0.10	**0.20**	0.30	0.50	0.65	0.80
Start dose	d			**60**				
Start	z	−1.47	−1.10	**−0.69**	−0.42	0.00	0.31	0.69
No DLT								
$y_1 = 0, q = 1.38$	z	−1.02	−0.73	−0.40	−0.17	0.21	0.50	0.87
Next dose	d_{NEXT}				**70**			
DLT observed								
$y_1 = 1, q = 0.72$	z	−2.07	−1.57	−1.06	−0.73	−0.24	0.10	0.51
Next dose	d_{NEXT}		**50**					

$y_1 = 0$ or 1). On the basis of this (single) observation, q (originally set equal to 1) is estimated.

If we assume the starting dose chosen such that Pr(DLT) $= \theta_{START} = 0.20$, then $d_1 = 60\,\text{mg/m}^2$ ($z = -0.69$). If, once a patient is tested at this dose, $y_1 = 1$, the next dose is lowered to $d_2 = 50$, whereas if $y_1 = 0$, it is raised to 70 (Table 3, Figs. 3, 4).

15.4 Practicalities

15.4.1 C33D or CRM

Although the CRM method is more efficient than the C33D design, it is considerably more difficult to implement, as the (statistical) manipulation required to determine the next dose to use is technically complex and requires specialist computer statistical software, although this is now freely available (Tan, Tan and Machin 2005).

Although this is more efficient than the "3+3" design, it is considerably more difficult to implement, as the (statistical) manipulation required is technically very complex. The design reduces the number of patients receiving the (very) low dose options. O'Quigley et al. (1990) argue that this avoids patients receiving doses at which there is little prospect of them driving benefit, but the design has been criticised by

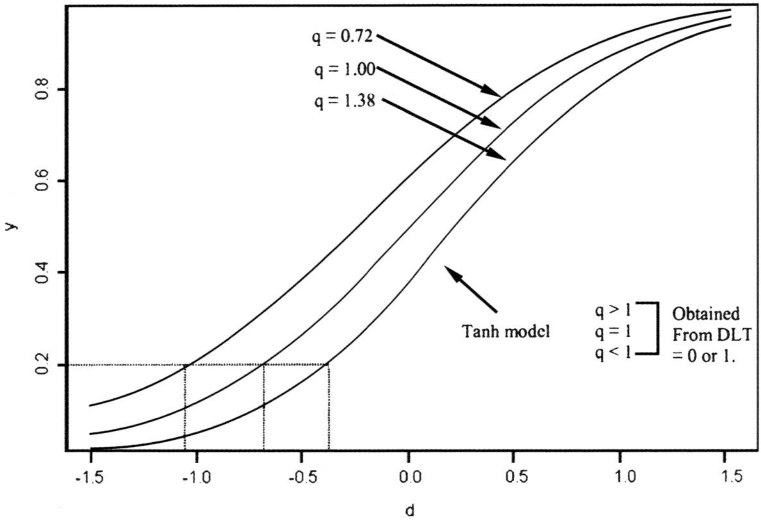

Fig. 3. Updated working dose response curves following a single observation investigating DLT ($y_1 = 0$ or 1) at first working dose $z = -0.69$. The results of the study by Flinn et al. (2000) are summarised in Fig. 4. At the close, they determine for patients with advanced NHL a MTD of daunoxome (with CVP) as 70 mg/m^2

Korn et al. (1994) for exposing patients to the risk of receiving potentially very toxic doses. However, modifications to the original design have been proposed to overcome both these difficulties (too low or too high). To our knowledge, few phase I trials have been published that have used the CRM design, although the workshop described by Eisenhauer et al. (2000) suggests this may not be the case. Further they recommend "... the approach of accruing three patients per dose level with dose escalation based on a modified Fibonacci sequence should no longer be considered the standard design."

However, the CRM procedures can be a somewhat of a black-box approach and this is not entirely satisfactory if the clinical (or statistical) teams involved do not understand the process. Thus Muler et al. (2004), when describing a phase I trial with a time-to-event outcome quote in their statistical estimation section state:

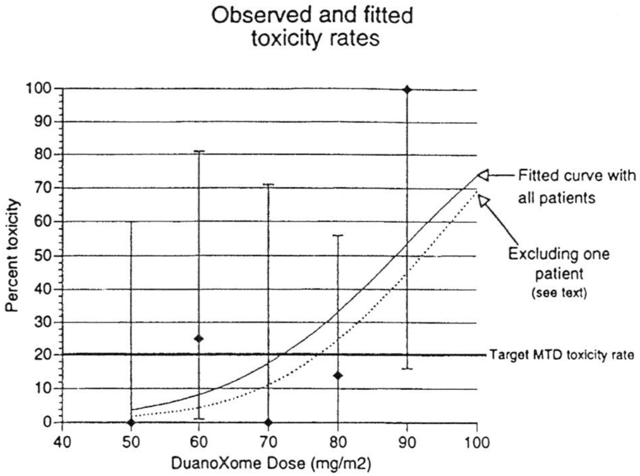

Fig. 4. Observed rates of DLT at each DaunoXome dose level. The point at which the curve crosses the 20% DLT line is the estimated MTD (from Flinn et al. 2000)

> At the end of the trial, the posterior distribution of the dose-toxicity parameter, α, and DLT at each dose, was calculated by means of Bayesian Analysis Using Gibbs Sampling software ... using the logistic dose-toxicity model and exponential prior distribution on α that were employed to conduct the trial. A burn-in of 40,000 iterations was followed by an estimation phase of 80,000 iterations, retaining every 10th value. Ninety-five percent posterior intervals for the toxicity probabilities were calculated from the sampled values.

Such phrasing offers no insight to the method used except for those very versed in such procedures.

It has to be recognised that phase I trials, however carefully designed, will include relatively few patients and so the corresponding level of uncertainty with respect to the true MTD will be high. It is also recognised that the designs do not (in one sense) estimate the MTD but rather choose one of the options presented by the investigators. This implies that very careful consideration needs to be given to the dose options available within the design.

One difficulty with some phase I designs is that the results from each patient must be known before the dose for the next patient can be determined. This almost certainly implies inbuilt delays in to the recruitment process and hence studies of lengthy duration.

Although phase I studies are often of modest or even small size, the temptation to conduct these studies without due attention to detail should be resisted. In fact, these studies (imprecise though they may be) provide key information for the drug development process. It is therefore essential that they are carefully designed, painstakingly conducted and meticulously reported in full.

Acknowledgements. Tan Sze-Huey and Loke Yee-Chong, National Cancer Centre, Singapore.

References

Bryant J, Day R (1995) Incorporating toxicity considerations into the design of two-stage phase II trials. Biometrics 51:1372–1383

Eisenhauer EA, O'Dwyer PJ, Christian M, Humphrey JS (2000) Phase I clinical trial design in cancer drug development. J Clin Oncol 18:684–692

Estlin EJ, Pinkerton CR, Lewis IJ et al (2001) A phase I study of nolatrexed dihydochloride in children with advanced cancer. A United Kingdom Children's Cancer Study Group Investigation. Br J Cancer 84:11–18

Flinn IW, Goodman SN, Post L, Jamison J, Miller CB, Gore S, Diehl L, Willis C, Ambinder RF, Byrd JC (2000) A dose-finding study of liposomal daunorubicin with CVP (COP-X) in advanced NHL. Ann Oncol 11:691–695

Korn EL, Midthune D, Chen TT, Rubinstein LV, Christian MC, Simon RM (1994) A comparison of two phase I trial designs. Stat Med 13:1799–1806

Muler JH, McGinn CJ, Normolle D, Lawrence T, Brown D, Hejna G, Zalupski MM (2004) Phase I trial using time-to-event continual reassessment strategy for dose escalation of cisplatin combined with gemcitabine and radiation therapy in pancreatic cancer. J Clin Oncol 22:238–243

O'Quigley J, Pepe M, Fisher L (1990) Continual reassessment method: a practical design for phase I trials. Biometrics 46:33–38

Smith M, Bernstein M, Bleyer WA, Borsi JD, Ho P, Lewis IJ, Pearson A, Pein F, Pratt C, Reaman G, Riccardi R, Seibel N, Trueworthy R, Ungerleider R, Vassal G, Vietti T (1998) Conduct of phase I trials in children with cancer. J Clin Oncol 16:966–978

Smith TL, Lee JJ, Kantarjian HM, Legha SS, Raber MN (1996) Design and results of phase I cancer clinical trials: three-year experience at M.D. Anderson Cancer Center. J Clin Oncol 14:287–295

Storer B (2001) Choosing a phase I design. In: Crowley J (ed) Handbook of statistics in clinical oncology. Marcel Dekker, New York, pp 73–91

Tan S-B, Tan S-H, Machin D (2005) Early phase clinical trials (EPCT) software, Version 1.0. National Cancer Centre, Singapore; UKCCSG, Leicester, UK and Clinical Trials and Epidemiology Research Unit (CTERU), Singapore

16 Preventing Postmarketing Changes in Recommended Doses and Marketing Withdrawals

C. Peck

Abstract. Recent market withdrawals of prescription drug products have brought attention to premarketing safety research. Less known but related to some drug withdrawals are postmarketing dosage changes of newly marketed drugs, including both dosage reductions and increases. These events have serious effects on patients, manufacturers, and regulatory authorities. Most of these harmful events could be avoided by intensive employment of targeted clinical pharmacology investigations to optimize dosage prior to phase III testing and regulatory approval. In this paper, the frequency and implications of postmarketing dosing changes and market withdrawals are considered in light of approaches to preventing them.

16.1 Introduction

Recent publicity concerning safety-motivated market withdrawals of prescription drugs has drawn both the pharmaceutical industry and regulatory agencies into the spotlight. A less publicized but closely related phenomenon is the official modification of recommended dosages of marketed prescription drugs. The need to improve premarketing dosing research to yield safer doses of approved therapeutic products was recognized by regulators over 30 years ago (Temple 1982) and has been frequently reinforced in the meantime (Temple 1989; International Conference on Harmonization 1994; Food and Drug Administration 2003a). In this paper, the frequency and implications of postmarketing dosing changes and market withdrawals are considered in light of approaches to preventing these costly events.

16.2 How Often Are Marketed Doses Changed?

"Historically, drugs have often been initially marketed at what were later recognized as excessive doses . . . sometimes with adverse consequences." (ICH 1994). This prophetic comment was recently confirmed in a study of all 499 new molecular entities approved by the FDA during the two decades spanning 1980–1999 (Cross et al. 2002). Comparing dosage recommendations in first approved labels with most recently approved labels up to the year 2002, these authors found that 20% of approved drugs underwent a 33% or greater labeled dosage modification, 80% of which were safety motivated dosage reductions (with no loss of effectiveness). Temporal dosage change rates, analyzed according to 5-year periods, showed ever faster dosage reductions following marketing, with drugs approved in 1995–1999 having a threefold rate relative to the 1980–1984 epoch. Independently, a European research team confirmed these dosage changes and trends using a different methodology, the WHO Defined Daily Dosage (Heerdink et al. 2002). Taken together, these studies confirm a systematic flaw in excess dosages selected and approved for marketing that appears to be recognized increasingly rapidly after regulatory approval.

16.3 Why Are Marketed Doses Changed?

While one in five marketed drugs underwent an increase in net dosage (increased loading or maintenance dose, dosing frequency or duration), such increases were usually based on observations of increased effectiveness at safe higher net doses: examples include octreotide, oxistat, butenafine, diltiazem, quinapril, mivacurium, cerivastatin, fluvastatin, indavir, irinotecan, and betaxol. A few dosage increases were motivated by increased convenience or diminished safety concern, e.g. ursodiol, granistetron, and risperidone.

Dosage decreases were of two magnitudes: (a) relabeled reductions in dosage, either reflected in the dosage section of the label or with a "black box warning" (severe contraindication due to concern about a very serious side effect) or (b) market withdrawals (dosage reduction to zero). In some cases, labeled dosage was decreased after recognition of the need for such changes in subpopulations with renal or hepatic impairment: cidofovir, troglitazone, tolcapone, and trovafloxacin. In other cases, metabolic drug interactions were cause for dosage reduction, e.g. sildenafil when ritonavir is also taken by the patient.

Attracting much negative attention to both the pharmaceutical industry and the regulatory authorities has been the highly publicized withdrawal of 13 drugs from the market since 1998: mibefradil, bromfenac, terfenadine, astemizole, grepafloxacin, alostron, cisapride, troglitazone, cerivastatin, rapacuronium, Vioxx, Tysari, and Bestra. FDA has defended the thesis that the drug withdrawal rate has been stable around 3% for many years (Food and Drug Administration 2003b). Since almost 50% of the recent withdrawals have been related to P450 drug metabolism-based drug–drug interactions and/or cardiotoxicity associated with QT-interval prolongation, the increasing clinical pharmacology information on these pharmacological attributes required by regulatory authorities is expected to minimize future withdrawals when safety is affected by these mechanisms. A recent report (Stolk et al. 2005) confirms that drugs that are substrates for P450 enzymes CYP2D and CYP3A are particularly vulnerable for postmarketing dosage change requirement, including high risk of market withdrawal.

16.4 What Are the Implications or Consequences of Getting the Marketed Dose Wrong?

Patients, pharmaceutical companies, and regulatory authorities are variably, but usually all affected by postmarketing dosage changes. Dosage increases may enable greater health benefits to patients and pharmaceutical sales, but some early patients may be denied the benefits of higher doses. Dosage decreases are usually stimulated by recognition of unacceptable toxicity in patients who sometimes sue the drug's manufacturer, publicity from which can decrease pharmaceutical sales. A safety motivated market withdrawal of a drug presents the most serious consequences to both patients and pharmaceutical companies. Not only are some patients harmed by excessive dosages, but future patients may be denied the benefits of a useful drug whose optimal dosage was never discovered. Lost revenues and class-action law suits can be threatening to the financial survival of a company. While regulatory authorities learn from dosage changes and withdrawals, such events often draw negative publicity and painful public hearings, which can prompt risk-averse regulatory reactions. In the case of Vioxx market removal, the FDA has initiated stricter premarketing safety requirements, as well as a multi-facetted patient-oriented initiative to improve the safety of marketed drugs in the future (see elements of the FDA's safety initiative below from the FDA website: http://www.fda.gov/cder/drugSafety.htm). When a massively marketed product such as Vioxx is withdrawn from the market, the entire pharmaceutical industry, accompanied by regulatory authorities may lose public confidence in drug products and the safeguards in place to assure safety.

FDA's New Drug Safety Initiative

FDA is launching a new program to make drug safety information available to you in an easily accessible format. Because patients are taking a more active role in their healthcare, we want to make safety information available about the medicines they are using. We believe that patients, their healthcare professionals, and other consumers will find the information we are providing useful in their prescribing and treatment decisions.

Our Drug Safety Initiative has the following components:

Drug safety information located together in a new web location

Drug specific information for healthcare professionals, patients, and other consumers
Other consumer education
Draft Guidance: FDA's "Drug Watch" for Emerging Drug Safety Information
Federal Register Notice of Availability – Draft Guidance for Industry on the FDA's "Drug Watch" for Emerging Drug Safety Information
Drug Safety Oversight Board
Questions and Answers on FDA's New Drug Safety Initiative

16.5 How to Avoid Postmarketing Dosage Changes

Four approaches are recommended to minimize and avoid postmarketing dosage changes.

– First, it is imperative to *secure a firm understanding of the clinical pharmacology and dose/exposure–response relationships (both efficacy and safety) in early development.* The FDA strongly encourages this approach in its various guidances on dose–response and exposure–response information (Food and Drug Administration 2003a, ICH 1994). The results of intensive clinical pharmacology learning in phase I and early phase II investigations can provide a firm basis for constructive dialogue with FDA in the newly established End of Phase IIa meeting (Food and Drug Administration 2003c) and the traditional End of Phase II meeting. This understanding will provide not only the information needed for more individualized dosing that promotes safety and effectiveness, but also more successful phase III trials, regulatory approval and stable marketed dosages.
– Second, *secure improved knowledge of predictable and possibly preventable adverse events prior to marketing authorization.* Aware that traditional effectiveness clinical trials involve relatively few subjects in a manner that is not reflective of real-life use, improved premarketing information on safety is needed. Recognizing the increasing ability to establish effectiveness efficiently, consideration can be given to employment of a Large, very Simple, Safety Trial (LvSST) (Peck 1999), undertaken prior to or after approval. Contemplated here is

a controlled trial involving 5,000–20,000 patients expected to benefit from the already-known-to-be effective new drug, prescribed under ordinary conditions of clinical practice. This safety-only trial would employ a very simple protocol that provides for documentation of only serious adverse reactions that are recorded on a highly simplified, minimal case report form. FDA has endorsed this concept in its report by the Task Force on Risk Management: "Wider use of large, community-based simple trials, restricting exposure during the early postmarketing period." (Food and Drug Administration 1999).

- Third, *undertake realistic education of marketers, prescribers and consumers*. Dispelling the myth that any drug is "safe," it is important to secure a common understanding that all drugs are inherently unsafe – except when the potential benefits outweigh the risks individually and collectively across all qualified patient candidates for the drug. Benefits of a drug must always be weighed against the known and potential risks of adverse reactions. Marketers must exert greater restraint on marketing to patients who are not candidates for the drug according to the indications in the label. Importantly, markers should curb prescriber incentive programs that draw prescribers into conflicts of interest that threaten their moral duty to protect their patients from harm. Prescribers must be patient advocates not only for access to needed drugs but to ensure informed matching of the right drug and dosage for eligible patients. Informed consumers who have a realistic understanding and expectations of safety can better modulate their therapeutic hopes and demands. Regulatory authorities must improve the labeling format and content so that labeled knowledge of drug safety is readable and understandable by the primary consumers of label information, e.g., prescribers and patients.
- Fourth, *manufacturers should make full use regulatory counsel and published guidance* that can mitigate costly development errors leading to postmarketing dosing changes. The ICH and FDA guidances cited above and closely related safety-oriented clinical pharmacology guidances on population pharmacokinetics (Food and Drug Administration 1999b), human drug metabolism, and drug–drug interactions (Food and Drug Aministration 1999c), pharmacokinetics in patients with impaired hepatic (Food and Drug Administration 2003d) and renal (Food and Drug Administration 1998) function, are must-read

scientific documents that reflect state-of-the-art knowledge. Direct communication with regulatory authorities on dosage optimization is increasingly available (Food and Drug Administration 2003c) and can be especially fruitful if manufacturers are researching critical information on factors influencing dose–exposure–response relationships for their new drug product.

16.6 Conclusions

A systematic flaw in dose-finding in contemporary drug development has led to postmarketing dosage changes in up to 20% of newly marketed drugs (mostly safety motivated dosage reductions), while 3% of all drugs have been withdrawn from the market due to unacceptable safety problems. These events have serious effects on patients (harm, deprivation of access to needed medications), manufacturers (litigation, lost revenues, battered reputation), and regulatory authorities (embarrassment, reactive risk-averse regulation). Many, if not most of these harmful events could be avoided by intensive employment of targeted clinical pharmacology investigations to optimize dosage prior to phase III testing and regulatory approval. A plethora of regulatory guidances are available to inform good dose-finding practices, which can be vetted via communications with regulatory authorities.

References

Cross J, Lee H, Westelinck A et al (2002) Postmarketing drug dosage changes of 499 FDA-approved new molecular entities, 1980–1999. Pharmacoepidemiol Drug Safety 11:439–446

Food and Drug Administration (1998) Guidance for industry pharmacokinetics in patients with impaired renal function – study design, data analysis, and impact on dosing and labeling. http://www.fda.gov/cder/guidance/1449fnl.pdf. Cited 6 April 2006

Food and Drug Administration (1999a) Managing risks from medical product use – creating a risk management framework. FDA Task Force on Risk Management. http://www.fda.gov/oc/tfrm/riskmanagement.pdf. Cited 6 April 2006

Food and Drug Administration (1999b) Guidance for industry – population pharmacokinetics. http://www.fda.gov/cder/guidance/1852fnl.pdf. Cited 6 April 2006

Food and Drug Administration (1999c) Guidance for industry – in vivo drug metabolism/drug interaction studies – study design, data analysis, and recommendations for dosing and labeling. http://www.fda.gov/cder/guidance/2635fnl.pdf. Cited 6 April 2006

Food and Drug Administration (2003a) Guidance for industry – exposure-response relationship – study design, data analysis, and regulatory applications. www.fda.gov/cder/guidance/index.htm. Cited 6 April 2006

Food and Drug Administration (2004b) Safety-based drug withdrawals in CDER Report to the Nation: 2004. http://www.fda.gov/cder/reports/rtn/2004/rtn2004–4.htm#Withdraw

Food and Drug Administration (2003c) Concept paper: end-of-phase-2A meetings with sponsors regarding exposure-response of IND and NDA Products. http://www.fda.gov/ohrms/dockets/ac/03/briefing/3998B1_01_Topic%201-Part%20A.pdf. Cited 6 April 2006

Food and Drug Administration (2003d) Guidance for industry pharmacokinetics in patients with impaired hepatic function: study design, data analysis, and impact on dosing and labeling. http://www.fda.gov/cder/guidance/3625fnl.pdf. Cited 6 April 2006

Heerdink E, Urquhart J, Leufkens HG (2002) Changes in prescribed drug doses after market introduction. Pharmacoepidemiol Drug Saf11:447–453

International Conference on Harmonization (1994) Guidance for industry: dose–response information to support drug registration. www.fda.gov/cder/guidance/ iche4.pdf

Peck C (1999) The large, *very simple*, safety trial (L*v*SST), presented at Safe Drugs at Any Speed? A Symposium on Drug Safety. (1999) Georgetown University, Washington DC, USA. personal communication – carl_peck@compuserve.com

Stolk P, Heerdink E, Leufkens HG (2005) Changes in the defined dialing dose; CYP2D6/CYP3A4 metabolism as an indicator for dose-setting problems. Eur J Clin Pharmacol 61:243–246

Temple R (1982) Government viewpoint of clinical trials. Drug Inf J 1982; 1: 10–17

Temple R (1989) Dose–response and registration of new drugs. In: Dose–response relationships in clinical pharmacology. Elsevier, Amsterdam

Ernst Schering Research Foundation Workshop

Editors: Günter Stock
Monika Lessl

Vol. 1 (1991): Bioscience ⇌ Societly Workshop Report
Editors: D.J. Roy, B.E. Wynne, R.W. Old

Vol. 2 (1991): Round Table Discussion on Bioscience ⇌ Society
Editor: J.J. Cherfas

Vol. 3 (1991): Excitatory Amino Acids and Second Messenger Systems
Editors: V.I. Teichberg, L. Turski

Vol. 4 (1992): Spermatogenesis – Fertilization – Contraception
Editors: E. Nieschlag, U.-F. Habenicht

Vol. 5 (1992): Sex Steroids and the Cardiovascular System
Editors: P. Ramwell, G. Rubanyi, E. Schillinger

Vol. 6 (1993): Transgenic Animals as Model Systems for Human Diseases
Editors: E.F. Wagner, F. Theuring

Vol. 7 (1993): Basic Mechanisms Controlling Term and Preterm Birth
Editors: K. Chwalisz, R.E. Garfield

Vol. 8 (1994): Health Care 2010
Editors: C. Bezold, K. Knabner

Vol. 9 (1994): Sex Steroids and Bone
Editors: R. Ziegler, J. Pfeilschifter, M. Bräutigam

Vol. 10 (1994): Nongenotoxic Carcinogenesis
Editors: A. Cockburn, L. Smith

Vol. 11 (1994): Cell Culture in Pharmaceutical Research
Editors: N.E. Fusenig, H. Graf

Vol. 12 (1994): Interactions Between Adjuvants, Agrochemical and Target Organisms
Editors: P.J. Holloway, R.T. Rees, D. Stock

Vol. 13 (1994): Assessment of the Use of Single Cytochrome P450 Enzymes in Drug Research
Editors: M.R. Waterman, M. Hildebrand

Vol. 14 (1995): Apoptosis in Hormone-Dependent Cancers
Editors: M. Tenniswood, H. Michna

Vol. 15 (1995): Computer Aided Drug Design in Industrial Research
Editors: E.C. Herrmann, R. Franke

Vol. 16 (1995): Organ-Selective Actions of Steroid Hormones
Editors: D.T. Baird, G. Schütz, R. Krattenmacher

Vol. 17 (1996): Alzheimer's Disease
Editors: J.D. Turner, K. Beyreuther, F. Theuring

Vol. 18 (1997): The Endometrium as a Target for Contraception
Editors: H.M. Beier, M.J.K. Harper, K. Chwalisz

Vol. 19 (1997): EGF Receptor in Tumor Growth and Progression
Editors: R.B. Lichtner, R.N. Harkins

Vol. 20 (1997): Cellular Therapy
Editors: H. Wekerle, H. Graf, J.D. Turner

Vol. 21 (1997): Nitric Oxide, Cytochromes P 450, and Sexual Steroid Hormones
Editors: J.R. Lancaster, J.F. Parkinson

Vol. 22 (1997): Impact of Molecular Biology and New Technical Developments in Diagnostic Imaging
Editors: W. Semmler, M. Schwaiger

Vol. 23 (1998): Excitatory Amino Acids
Editors: P.H. Seeburg, I. Bresink, L. Turski

Vol. 24 (1998): Molecular Basis of Sex Hormone Receptor Function
Editors: H. Gronemeyer, U. Fuhrmann, K. Parczyk

Vol. 25 (1998): Novel Approaches to Treatment of Osteoporosis
Editors: R.G.G. Russell, T.M. Skerry, U. Kollenkirchen

Vol. 26 (1998): Recent Trends in Molecular Recognition
Editors: F. Diederich, H. Künzer

Vol. 27 (1998): Gene Therapy
Editors: R.E. Sobol, K.J. Scanlon, E. Nestaas, T. Strohmeyer

Vol. 28 (1999): Therapeutic Angiogenesis
Editors: J.A. Dormandy, W.P. Dole, G.M. Rubanyi

Vol. 29 (2000): Of Fish, Fly, Worm and Man
Editors: C. Nüsslein-Volhard, J. Krätzschmar

Vol. 30 (2000): Therapeutic Vaccination Therapy
Editors: P. Walden, W. Sterry, H. Hennekes

Vol. 31 (2000): Advances in Eicosanoid Research
Editors: C.N. Serhan, H.D. Perez

Vol. 32 (2000): The Role of Natural Products in Drug Discovery
Editors: J. Mulzer, R. Bohlmann

Vol. 33 (2001): Stem Cells from Cord Blood, In Utero Stem Cell Development, and Transplantation-Inclusive Gene Therapy
Editors: W. Holzgreve, M. Lessl

Vol. 34 (2001): Data Mining in Structural Biology
Editors: I. Schlichting, U. Egner

Vol. 35 (2002): Stem Cell Transplantation and Tissue Engineering
Editors: A. Haverich, H. Graf

Vol. 36 (2002): The Human Genome
Editors: A. Rosenthal, L. Vakalopoulou

Vol. 37 (2002): Pharmacokinetic Challenges in Drug Discovery
Editors: O. Pelkonen, A. Baumann, A. Reichel

Vol. 38 (2002): Bioinformatics and Genome Analysis
Editors: H.-W. Mewes, B. Weiss, H. Seidel

Vol. 39 (2002): Neuroinflammation – From Bench to Bedside
Editors: H. Kettenmann, G.A. Burton, U. Moenning

Vol. 40 (2002): Recent Advances in Glucocorticoid Receptor Action
Editors: A. Cato, H. Schaecke, K. Asadullah

Vol. 41 (2002): The Future of the Oocyte
Editors: J. Eppig, C. Hegele-Hartung

Vol. 42 (2003): Small Molecule-Protein Interaction
Editors: H. Waldmann, M. Koppitz

Vol. 43 (2003): Human Gene Therapy:
Present Opportunities and Future Trends
Editors: G.M. Rubanyi, S. Ylä-Herttuala

Vol. 44 (2004): Leucocyte Trafficking:
The Role of Fucosyltransferases and Selectins
Editors: A. Hamann, K. Asadullah, A. Schottelius

Vol. 45 (2004): Chemokine Roles in Immunoregulation and Disease
Editors: P.M. Murphy, R. Horuk

Vol. 46 (2004): New Molecular Mechanisms of Estrogen Action and Their Impact on Future Perspectives in Estrogen Therapy
Editors: K.S. Korach, A. Hillisch, K.H. Fritzemeier

Vol. 47 (2004): Neuroinflammation in Stroke
Editors: U. Dirnagl, B. Elger

Vol. 48 (2004): From Morphological Imaging to Molecular Targeting
Editors: M. Schwaiger, L. Dinkelborg, H. Schweinfurth

Vol. 49 (2004): Molecular Imaging
Editors: A.A. Bogdanov, K. Licha

Vol. 50 (2005): Animal Models of T Cell-Mediated Skin Diseases
Editors: T. Zollner, H. Renz, K. Asadullah

Vol. 51 (2005): Biocombinatorial Approaches for Drug Finding
Editors: W. Wohlleben, T. Spellig, B. Müller-Tiemann

Vol. 52 (2005): New Mechanisms for Tissue-Selective Estrogen-Free Contraception
Editors: H.B. Croxatto, R. Schürmann, U. Fuhrmann, I. Schellschmidt

Vol. 53 (2005): Opportunities and Challenges of the Therapies Targeting CNS Regeneration
Editors: D. Perez, B. Mitrovic, A. Baron Van Evercooren

Vol. 54 (2005): The Promises and Challenges of Regenerative Medicine
Editors: J. Morser, S.I. Nishikawa

Vol. 55 (2006): Chronic Viral and Inflammatory Cardiomyopathy
Editors: H.-P. Schultheiss, J.-F. Kapp

Vol. 56 (2006): Cytokines as Potential Therapeutic Target for Inflammatory Skin Diseases
Editors: R. Numerof, C.A. Dinarello, K. Asadullah

Vol. 57 (2006): The Histone Code and Beyond
Editors: S.L. Berger, O. Nakanishi, B. Haendler

Vol. 58 (2006): Chemical Genomics
Editors: S. Jaroch, H. Weinmann

Vol. 59 (2006): Appropriate Dose Selection – How to Optimize Clinical Drug Development
Editors: J. Venitz, W. Sittner

Supplement 1 (1994): Molecular and Cellular Endocrinology of the Testis
Editors: G. Verhoeven, U.-F. Habenicht

Supplement 2 (1997): Signal Transduction in Testicular Cells
Editors: V. Hansson, F.O. Levy, K. Taskén

Supplement 3 (1998): Testicular Function:
From Gene Expression to Genetic Manipulation
Editors: M. Stefanini, C. Boitani, M. Galdieri, R. Geremia,F. Palombi

Supplement 4 (2000): Hormone Replacement Therapy
and Osteoporosis
Editors: J. Kato, H. Minaguchi, Y. Nishino

Supplement 5 (1999): Interferon: The Dawn of Recombinant
Protein Drugs
Editors: J. Lindenmann, W.D. Schleuning

Supplement 6 (2000): Testis, Epididymis and Technologies
in the Year 2000
Editors: B. Jégou, C. Pineau, J. Saez

Supplement 7 (2001): New Concepts in Pathology
and Treatment of Autoimmune Disorders
Editors: P. Pozzilli, C. Pozzilli, J.-F. Kapp

Supplement 8 (2001): New Pharmacological Approaches
to Reproductive Health and Healthy Ageing
Editors: W.-K. Raff, M.F. Fathalla, F. Saad

Supplement 9 (2002): Testicular Tangrams
Editors: F.F.G. Rommerts, K.J. Teerds

Supplement 10 (2002): Die Architektur des Lebens
Editors: G. Stock, M. Lessl

Supplement 11 (2005): Regenerative and Cell Therapy
Editors: A. Keating, K. Dicke, N. Gorin, R. Weber, H. Graf

Supplement 12 (2005): Von der Wahrnehmung zur Erkenntnis –
From Perception to Understanding
Editors: M. Lessl, J. Mittelstraß

Supplement 13 (2006): Biologie und Epidemiologie
der Hormonersatztherapie –
Biology and Epidemiology of Hormone Replacement Therapy
Editors: M.A. Lewis, M. Dietel, P.C. Scriba, W.K. Raff